INTERNATIONAL CODES®

2018 I-CODE BONUS OFFER

Get a **free 45-day online subscription** to ICC's *premiumACCESS*™ 2018 I-Codes Complete Collection. Test drive many powerful, time-saving tools available to you from *premiumACCESS*. To activate your bonus, visit **www.iccsafe.org/codebonus**.

IPC®

A Member of the International Code Family®

INTERNATIONAL PLUMBING CODE®

2018 International Plumbing Code®

Date of First Publication: August 31, 2017

First Printing: August 2017
Second Printing: February 2018
Third Printing: September 2019

ISBN: 978-1-60983-745-7 (soft-cover edition)
ISBN: 978-1-60983-744-0 (loose-leaf edition)

COPYRIGHT © 2017
by
INTERNATIONAL CODE COUNCIL, INC.

ALL RIGHTS RESERVED. This 2018 *International Plumbing Code®* is a copyrighted work owned by the International Code Council, Inc. ("ICC"). Without advance written permission from the ICC, no part of this book may be reproduced, distributed or transmitted in any form or by any means, including, without limitation, electronic, optical or mechanical means (by way of example, and not limitation, photocopying, or recording by or in an information storage retrieval system). For information on use rights and permissions, please contact: ICC Publications, 4051 Flossmoor Road, Country Club Hills, IL 60478. Phone 1-888-ICC-SAFE (422-7233).

Trademarks: "International Code Council," the "International Code Council" logo, "ICC," the "ICC" logo, "International Plumbing Code," "IPC" and other names and trademarks appearing in this book are registered trademarks of the International Code Council, Inc., and/or its licensors (as applicable), and may not be used without permission.

The display of the Plumbing-Heating-Cooling Contractors—National Association (PHCC) logo in this publication indicates PHCC's support through committee participation of ICC's open governmental consensus process used to develop the International Codes. This support does not imply any ownership to the copyright to the *International Plumbing Code*, which is held solely by the International Code Council, Inc.

Plumbing-Heating-Cooling Contractors—National Association (PHCC): 180 S. Washington Street – Suite 100, Falls Church, VA 22046; Phone: (703) 237-8100, (800) 533-7694, www.phccweb.org.

PRINTED IN THE USA

PREFACE

Introduction

The *International Plumbing Code*® (IPC®) establishes minimum requirements for plumbing systems using prescriptive and performance-related provisions. It is founded on broad-based principles that make possible the use of new materials and new plumbing designs. This 2018 edition is fully compatible with all of the *International Codes*® (I-Codes®) published by the International Code Council® (ICC®), including the *International Building Code*®, *International Energy Conservation Code*®, *International Existing Building Code*®, *International Fire Code*®, *International Fuel Gas Code*®, *International Green Construction Code*®, *International Mechanical Code*®, *International Private Sewage Disposal Code*®, *International Property Maintenance Code*®, *International Residential Code*®, *International Swimming Pool and Spa Code*®, *International Wildland-Urban Interface Code*®, *International Zoning Code*® and *International Code Council Performance Code*®.

The I-Codes, including this *International Plumbing Code*, are used in a variety of ways in both the public and private sectors. Most industry professionals are familiar with the I-Codes as the basis of laws and regulations in communities across the U.S. and in other countries. However, the impact of the codes extends well beyond the regulatory arena, as they are used in a variety of nonregulatory settings, including:

- Voluntary compliance programs such as those promoting sustainability, energy efficiency and disaster resistance.
- The insurance industry, to estimate and manage risk, and as a tool in underwriting and rate decisions.
- Certification and credentialing of individuals involved in the fields of building design, construction and safety.
- Certification of building and construction-related products.
- U.S. federal agencies, to guide construction in an array of government-owned properties.
- Facilities management.
- "Best practices" benchmarks for designers and builders, including those who are engaged in projects in jurisdictions that do not have a formal regulatory system or a governmental enforcement mechanism.
- College, university and professional school textbooks and curricula.
- Reference works related to building design and construction.

In addition to the codes themselves, the code development process brings together building professionals on a regular basis. It provides an international forum for discussion and deliberation about building design, construction methods, safety, performance requirements, technological advances and innovative products.

Development

This 2018 edition presents the code as originally issued, with changes reflected in the 2003 through 2015 editions and with further changes approved by the ICC Code Development Process through 2017. A new edition such as this is promulgated every 3 years.

This code is founded on principles intended to establish provisions consistent with the scope of a plumbing code that adequately protects public health, safety and welfare; provisions that do not unnecessarily increase construction costs; provisions that do not restrict the use of new materials, products or methods of construction; and provisions that do not give preferential treatment to particular types or classes of materials, products or methods of construction.

Maintenance

The *International Plumbing Code* is kept up to date through the review of proposed changes submitted by code enforcement officials, industry representatives, design professionals and other interested parties. Proposed changes are carefully considered through an open code development process in which all interested and affected parties may participate.

The ICC Code Development Process reflects principles of openness, transparency, balance, due process and consensus, the principles embodied in OMB Circular A-119, which governs the federal government's use of private-sector standards. The ICC process is open to anyone; there is no cost to participate, and people can participate without travel cost through the ICC's cloud-based app, cdpAccess®. A broad cross section of interests are represented in the ICC Code Development Process. The codes, which are updated regularly, include safeguards that allow for emergency action when required for health and safety reasons.

In order to ensure that organizations with a direct and material interest in the codes have a voice in the process, the ICC has developed partnerships with key industry segments that support the ICC's important public safety mission. Some code development committee members were nominated by the following industry partners and approved by the ICC Board:

- American Institute of Architects (AIA)
- American Society of Plumbing Engineers (ASPE)
- National Association of Home Builders (NAHB)
- Plumbing Heating and Cooling Contractors (PHCC)

The Code Development Committees evaluate and make recommendations regarding proposed changes to the codes. Their recommendations are then subject to public comment and council-wide votes. The ICC's governmental members—public safety officials who have no financial or business interest in the outcome—cast the final votes on proposed changes.

The contents of this work are subject to change through the code development cycles and by any governmental entity that enacts the code into law. For more information regarding the code development process, contact the Codes and Standards Development Department of the International Code Council.

While the I-Code development procedure is thorough and comprehensive, the ICC, its members and those participating in the development of the codes disclaim any liability resulting from the publication or use of the I-Codes, or from compliance or noncompliance with their provisions. The ICC does not have the power or authority to police or enforce compliance with the contents of this code.

Code Development Committee Responsibilities (Letter Designations in Front of Section Numbers)

In each code development cycle, proposed changes to the code are considered at the Committee Action Hearings by the International Plumbing Code Development Committee, whose action constitutes a recommendation to the voting membership for final action on the proposed change. Proposed changes to a code section that has a number beginning with a letter in brackets are considered by a different code development committee. For example, proposed changes to code sections that have [BS] in front of them (e.g., [BS] 309.2) are considered by the IBC—Structural Code Development Committee at the code development hearings.

The bracketed letter designations for committees responsible for portions of this code are as follows:

[A] = Administrative Code Development Committee;

[BE] = IBC—Egress Code Development Committee;

[BG] = IBC—General Code Development Committee;

[BS] = IBC—Structural Code Development Committee;

[E] = International Energy Conservation Code Development Committee;

[F] = International Fire Code Development Committee; and

[M] = International Mechanical Code Development Committee.

For the development of the 2021 edition of the I-Codes, there will be two groups of code development committees and they will meet in separate years. Note that these are tentative groupings.

Group A Codes (Heard in 2018, Code Change Proposals Deadline: January 8, 2018)	Group B Codes (Heard in 2019, Code Change Proposals Deadline: January 7, 2019)
International Building Code – Egress (Chapters 10, 11, Appendix E) – Fire Safety (Chapters 7, 8, 9, 14, 26) – General (Chapters 2–6, 12, 27–33, Appendices A, B, C, D, K, N)	Administrative Provisions (Chapter 1 of all codes except IECC, IRC and IgCC, administrative updates to currently referenced standards, and designated definitions)
International Fire Code	**International Building Code** – Structural (Chapters 15–25, Appendices F, G, H, I, J, L, M)
International Fuel Gas Code	**International Existing Building Code**
International Mechanical Code	**International Energy Conservation Code—Commercial**
International Plumbing Code	**International Energy Conservation Code—Residential** – IECC—Residential – IRC—Energy (Chapter 11)
International Property Maintenance Code	**International Green Construction Code** (Chapter 1)
International Private Sewage Disposal Code	**International Residential Code** – IRC—Building (Chapters 1–10, Appendices E, F, H, J, K, L, M, O, Q, R, S, T)
International Residential Code – IRC—Mechanical (Chapters 12–23) – IRC—Plumbing (Chapters 25–33, Appendices G, I, N, P)	
International Swimming Pool and Spa Code	
International Wildland-Urban Interface Code	
International Zoning Code	
Note: Proposed changes to the ICC *Performance Code*™ will be heard by the code development committee noted in brackets [ ] in the text of the ICC *Performance Code*™.	

Code change proposals submitted for code sections that have a letter designation in front of them will be heard by the respective committee responsible for such code sections. Because different committees hold Committee Action Hearings in different years, proposals for the IPC will be heard by committees in both the 2018 (Group A) and the 2019 (Group B) code development cycles.

For instance, every section of Chapter 1 of this code is designated as the responsibility of the Administrative Code Development Committee, which is part of the Group B portion of the hearings. This committee will hold its Committee Action Hearings in 2019 to consider code change proposals for Chapter 1 of all I-Codes except the *International Energy Conservation Code*, *International Residential Code* and *International Green Construction Code*. Therefore, any proposals received for Chapter 1 of this code will be assigned to the Administrative Code Development Committee for consideration in 2019.

It is very important that anyone submitting code change proposals understands which code development committee is responsible for the section of the code that is the subject of the code change proposal. For further information on the Code Development Committee responsibilities, please visit the ICC website at www.iccsafe.org/scoping.

Marginal Markings

Solid vertical lines in the margins within the body of the code indicate a technical change from the requirements of the 2015 edition. Deletion indicators in the form of an arrow (➡) are provided in the margin where an entire section, paragraph, exception or table has been deleted or an item in a list of items or a table has been deleted.

A single asterisk [*] placed in the margin indicates that text or a table has been relocated within the code. A double asterisk [**] placed in the margin indicates that the text or table immediately following it has been relocated there from elsewhere in the code. The following table indicates such relocations in the 2018 edition of the *International Plumbing Code*.

2018 LOCATION	2015 LOCATION
802.2	804.1

Coordination of the International Codes

The coordination of technical provisions is one of the strengths of the ICC family of model codes. The codes can be used as a complete set of complementary documents, which will provide users with full integration and coordination of technical provisions. Individual codes can also be used in subsets or as stand-alone documents. To make sure that each individual code is as complete as possible, some technical provisions that are relevant to more than one subject area are duplicated in some of the model codes. This allows users maximum flexibility in their application of the I-Codes.

Italicized Terms

Words and terms defined in Chapter 2, Definitions, are italicized where they appear in code text and the Chapter 2 definition applies. Where such words and terms are not italicized, common-use definitions apply. The words and terms selected have code-specific definitions that the user should read carefully to facilitate better understanding of the code.

Adoption

The International Code Council maintains a copyright in all of its codes and standards. Maintaining copyright allows the ICC to fund its mission through sales of books, in both print and electronic formats. The ICC welcomes adoption of its codes by jurisdictions that recognize and acknowledge the ICC's copyright in the code, and further acknowledge the substantial shared value of the public/private partnership for code development between jurisdictions and the ICC.

The ICC also recognizes the need for jurisdictions to make laws available to the public. All I-Codes and I-Standards, along with the laws of many jurisdictions, are available for free in a nondownloadable form on the ICC's website. Jurisdictions should contact the ICC at adoptions@iccsafe.org to learn how to adopt and distribute laws based on the *International Plumbing Code* in a manner that provides necessary access, while maintaining the ICC's copyright.

To facilitate adoption, several sections of this code contain blanks for fill-in information that needs to be supplied by the adopting jurisdiction as part of the adoption legislation. For this code, please see:

Section 101.1. Insert: [NAME OF JURISDICTION]

Section 106.6.2. Insert: [APPROPRIATE SCHEDULE]

Section 106.6.3. Insert: [PERCENTAGES IN TWO LOCATIONS]

Section 108.4. Insert: [OFFENSE, DOLLAR AMOUNT, NUMBER OF DAYS]

Section 108.5. Insert: [DOLLAR AMOUNT IN TWO LOCATIONS]

Section 305.4.1. Insert: [NUMBER OF INCHES IN TWO LOCATIONS]

Section 903.1. Insert: [NUMBER OF INCHES]

EFFECTIVE USE OF THE INTERNATIONAL PLUMBING CODE

The *International Plumbing Code* (IPC) is a model code that regulates the design and installation of plumbing systems including the plumbing fixtures in all types of buildings except for detached one- and two-family dwellings and townhouses that are not more than three stories above grade in height. The regulations for plumbing systems in one- and two-family dwellings and townhouses are covered by Chapters 25 through 33 of the *International Residential Code* (IRC). The IPC addresses general plumbing regulations, fixture requirements, water heater installations and systems for water distribution, sanitary drainage, special wastes, venting, storm drainage and medical gases. The IPC does not address fuel gas piping systems as those systems are covered by the *International Fuel Gas Code* (IFGC). The IPC also does not regulate swimming pool piping systems, process piping systems, or utility-owned piping and systems. The purpose of the IPC is to the establish the minimum acceptable level of safety to protect life and property from the potential dangers associated with supplying potable water to plumbing fixtures and outlets and the conveyance of bacteria-laden wastewater from fixtures.

The IPC is primarily a specification-oriented (prescriptive) code with some performance-oriented text. For example, Section 405.1 is a performance statement but Chapter 6 contains the prescriptive requirements that will cause Section 405.1 to be satisfied.

Where a building contains plumbing fixtures, those fixtures requiring water must be provided with an adequate supply of water for proper operation. The number of required plumbing fixtures for a building is specified by this code and is based upon the anticipated maximum number of occupants for the building and the type of building occupancy. This code provides prescriptive criteria for sizing piping systems connected to those fixtures. Through the use of code-approved materials and the installation requirements specified in this code, plumbing systems will perform their intended function over the life of the building. In summary, the IPC sets forth the minimum requirements for providing safe water to a building as well as a safe manner in which liquid-borne wastes are carried away from a building.

Arrangement and Format of the 2018 IPC

The format of the IPC allows each chapter to be devoted to a particular subject with the exception of Chapter 3 which contains general subject matters that are not extensive enough to warrant their own independent chapter.

Chapters	Subjects
1–2	Administration and Definitions
3	General Regulations
4	Fixtures, Faucets and Fixture Fittings
5	Water Heaters
6	Water Supply and Distribution
7	Sanitary Drainage
8	Indirect/Special Waste
9	Vents
10	Traps, Interceptors and Separators
11	Storm Drainage
12	Special Piping (Medical Gas)
13	Nonpotable Water Systems
14	Subsurface Landscape Irrigation Systems
15	Referenced Standards
Appendices A–E	Appendices

The following is a chapter-by-chapter synopsis of the scope and intent of the provisions of the *International Plumbing Code*:

Chapter 1 Scope and Administration. This chapter contains provisions for the application, enforcement and administration of subsequent requirements of the code. In addition to establishing the scope of the code, Chapter 1 identifies which buildings and structures come under its purview. Chapter 1 is largely concerned with maintaining "due process of law" in enforcing the requirements contained in the body of this code. Only through careful observation of the administrative provisions can the code official reasonably expect to demonstrate that "equal protection under the law" has been provided.

Chapter 2 Definitions. Chapter 2 is the repository of the definitions of terms used in the body of the code. Codes are technical documents and every word, term and punctuation mark can impact the meaning of the code text and the intended results. The code often uses terms that have a unique meaning in the code and the code meaning can differ substantially from the ordinarily understood meaning of the term as used outside of the code.

The terms defined in Chapter 2 are deemed to be of prime importance in establishing the meaning and intent of the code text that uses the terms. The user of the code should be familiar with and consult this chapter because the definitions are essential to the correct interpretation of the code and because the user may not be aware that a term is defined.

Where understanding of a term's definition is especially key to or necessary for understanding of a particular code provision, the term is shown in *italics*. This is true only for those terms that have a meaning that is unique to the code. In other words, the generally understood meaning of a term or phrase might not be sufficient or consistent with the meaning prescribed by the code; therefore, it is essential that the code-defined meaning be known.

Guidance regarding tense, gender and plurality of defined terms as well as guidance regarding terms not defined in this code is provided.

Chapter 3 General Regulations. The content of Chapter 3 is often referred to as "miscellaneous," rather than general regulations. This is the only chapter in the code whose requirements do not interrelate. If a requirement cannot be located in another chapter, it should be located in this chapter. Chapter 3 contains safety requirements for the installation of plumbing and nonplumbing requirements for all types of fixtures. This chapter also has requirements for the identification of pipe, pipe fittings, traps, fixtures, materials and devices used in plumbing systems.

The safety requirements of this chapter provide protection for the building's structural members, as well as prevent undue stress and strain on pipes. The building's structural stability is protected by the regulations for cutting and notching of structural members. Additional protection for the building occupants includes requirements to maintain the plumbing in a safe and sanitary condition, as well as privacy for those occupants.

Chapter 4 Fixtures, Faucets and Fixture Fittings. This chapter regulates the minimum number of plumbing fixtures that must be provided for every type of building. This chapter also regulates the quality of fixtures and faucets by requiring those items to comply with nationally recognized standards. Because fixtures must be properly installed so that they are usable by the occupants of the building, this chapter contains the requirements for the installation of fixtures. Because the requirements for the number of plumbing fixtures affects the design of a building, Chapter 29 of the *International Building Code* (IBC) includes, verbatim, many of the requirements listed in Chapter 4 of this code.

Chapter 5 Water Heaters. Chapter 5 regulates the design, approval and installation of water heaters and related safety devices. The intent is to minimize the hazards associated with the installation and operation of water heaters. Although this code does not regulate the size of a water heater, it does regulate all other aspects of the water heater installation such as temperature and pressure relief valves, safety drip pans, installation and connections. Where a water heater also supplies water for space heating, this chapter regulates the maximum water temperature supplied to the water distribution system.

Chapter 6 Water Supply and Distribution. This chapter regulates the supply of potable water from both public and individual sources to every fixture and outlet so that it remains potable and uncontaminated. Chapter 6 also regulates the design of the water distribution system, which will allow fixtures to function properly and also help prevent backflow conditions. The unique requirements of the water supply for health care facilities are addressed separately. It is critical that the potable water supply system remain free of actual or potential sanitary hazards by providing protection against backflow.

Chapter 7 Sanitary Drainage. The purpose of Chapter 7 is to regulate the materials, design and installation of sanitary drainage piping systems as well as the connections made to the system. The intent is to design and install sanitary drainage systems that will function reliably, that are neither undersized nor oversized and that are constructed from materials, fittings and connections as prescribed herein. This chapter addresses the proper use of fittings for directing the flow into and within the sanitary drain piping system. Materials and provisions necessary for servicing the drainage system are also included in this chapter.

Chapter 8 Indirect/Special Waste. This chapter regulates drainage installations that require an indirect connection to the sanitary drainage system. Fixtures and plumbing appliances, such as those associated with food preparation or handling, health care facilities and potable liquids, must be protected from contamination that can result from connection to the drainage system. An indirect connection prevents sewage from backing up into a fixture or appliance, thus providing protection against potential health hazards. The chapter also regulates special wastes containing hazardous chemicals. Special waste must be treated to prevent any damage to the sanitary drainage piping and to protect the sewage treatment processes.

Chapter 9 Vents. Chapter 9 covers the requirements for vents and venting. Knowing why venting is required makes it easier to understand the intent of this chapter. Venting protects every trap against the loss of its seal. Provisions set forth in this chapter are geared toward limiting the pressure differentials in the drainage system to a maximum of 1 inch of water column (249 Pa) above or below atmospheric pressure (i.e., positive or negative pressures).

Chapter 10 Traps, Interceptors and Separators. This chapter contains design requirements and installation limitations for traps. Prohibited types of traps are specifically identified. Where fixtures do not frequently replenish the water in traps, a method is provided to ensure that the water seal of the trap will be maintained. Requirements for the design and location of various types of interceptors and separators are provided. Specific venting requirements are given for separators and interceptors as those requirements are not addressed in Chapter 9.

Chapter 11 Storm Drainage. Chapter 11 regulates the removal of storm water typically associated with rainfall. The proper installation of a storm drainage system reduces the possibility of structural collapse of a flat roof, prevents the leakage of water through the roof, prevents damage to the footings and foundation of the building and prevents flooding of the lower levels of the building.

Chapter 12 Special Piping and Storage Systems. This chapter contains the requirements for the design, installation, storage, handling and use of nonflammable medical gas systems, including inhalation anesthetic and vacuum piping systems, bulk oxygen storage systems and oxygen-fuel gas systems used for welding and cutting operations. The intent of these requirements is to minimize the potential fire and explosion hazards associated with the gases used in these systems.

Chapter 13 Nonpotable Water Systems. This chapter regulates the design and installation of nonpotable water systems. The reduction of potable water use in buildings has led building designers in some jurisdictions to use nonpotable water for irrigation and flushing of water closets and urinals. This chapter provides the overall requirements for these systems.

Chapter 14 Subsurface Landscape Irrigation Systems. This chapter regulates the design and installation of subsurface landscape irrigation systems for the disposal of on-site nonpotable water such as graywater. The reduction of potable water use in buildings has led building designers in some jurisdictions to use on-site nonpotable water for irrigation. This chapter provides the overall requirements for these systems.

Chapter 15 Referenced Standards. Chapter 15 contains a comprehensive list of all standards that are referenced in the code. The standards are part of the code to the extent of the reference to the standard. Compliance with the referenced standard is necessary for compliance with this code. By providing specifically adopted standards, the construction and installation requirements necessary for compliance with the code can be readily determined. The basis for code compliance is, therefore, established and available on an equal basis to the code official, contractor, designer and owner.

Chapter 15 is organized in a manner that makes it easy to locate specific standards. It lists all of the referenced standards, alphabetically, by acronym of the promulgating agency of the standard. Each agency's standards are then listed in either alphabetical or numeric order based upon the standard identification. The list also contains the title of the standard; the edition (date) of the standard referenced; any addenda included as part of the ICC adoption; and the section or sections of this code that reference the standard.

Appendix A Plumbing Permit Fee Schedule. Appendix A provides a format for a fee schedule.

Appendix B Rates of Rainfall for Various Cities. Appendix B provides specific rainfall rates for major cities in the United States.

Appendix C Structural Safety. Appendix C is provided so that the user does not have to refer to another code book for limitations for cutting, notching and boring of sawn lumber and cold-formed steel framing.

Appendix D Degree Day and Design Temperatures. This appendix provides valuable temperature information for designers and installers of plumbing systems in areas where freezing temperatures might exist.

Appendix E Sizing of Water Piping System. Appendix E provides two recognized methods for sizing the water service and water distribution piping for any structure. The method under Section E103 provides friction loss diagrams which require the user to "plot" points and read values from the diagrams in order to perform the required calculations and necessary checks. This method is the most accurate of the two presented in this appendix. The method under Section E201 is known to be conservative; however, very few calculations are necessary in order to determine a pipe size that satisfies the flow requirements of any application.

TABLE OF CONTENTS

CHAPTER 1	**SCOPE AND ADMINISTRATION**	**1**

PART 1—SCOPE AND APPLICATION............ 1
Section
101 General.................................... 1
102 Applicability................................ 1

PART 2—ADMINISTRATION AND ENFORCEMENT................. 2
Section
103 Department of Plumbing Inspection............. 2
104 Duties and Powers of the Code Official.......... 2
105 Approval.................................... 3
106 Permits..................................... 3
107 Inspections and Testing....................... 5
108 Violations................................... 7
109 Means of Appeal............................. 8
110 Temporary Equipment, Systems and Uses....... 8

CHAPTER 2   DEFINITIONS..................... 9
Section
201 General.................................... 9
202 General Definitions.......................... 9

CHAPTER 3   GENERAL REGULATIONS........ 17
Section
301 General................................... 17
302 Exclusion of Materials Detrimental to the Sewer System....................... 17
303 Materials.................................. 17
304 Rodentproofing............................. 18
305 Protection of Pipes and Plumbing System Components...................... 18
306 Trenching, Excavation and Backfill............ 18
307 Structural Safety............................ 19
308 Piping Support.............................. 19
309 Flood Hazard Resistance..................... 20
310 Washroom and Toilet Room Requirements...... 20
311 Toilet Facilities for Workers.................. 21
312 Tests and Inspections....................... 21
313 Equipment Efficiencies...................... 22
314 Condensate Disposal........................ 22
315 Penetrations............................... 23
316 Alternative Engineered Design................ 23

CHAPTER 4   FIXTURES, FAUCETS AND FIXTURE FITTINGS......... 25
Section
401 General................................... 25
402 Fixture Materials........................... 25
403 Minimum Plumbing Facilities................. 25
404 Accessible Plumbing Facilities................ 29
405 Installation of Fixtures....................... 29
406 Automatic Clothes Washers.................. 30
407 Bathtubs................................... 30
408 Bidets..................................... 30
409 Dishwashing Machines...................... 31
410 Drinking Fountains.......................... 31
411 Emergency Showers and Eyewash Stations...... 31
412 Faucets and other Fixture Fittings............. 31
413 Floor and Trench Drains..................... 32
414 Floor Sinks................................ 32
415 Flushing Devices for Water Closets and Urinals... 32
416 Food Waste Disposer Units................... 33
417 Garbage Can Washers....................... 33
418 Laundry Trays.............................. 33
419 Lavatories................................. 33
420 Manual Food and Beverage Dispensing Equipment.................... 33
421 Showers................................... 33
422 Sinks..................................... 34
423 Specialty Plumbing Fixtures.................. 35
424 Urinals.................................... 35
425 Water Closets.............................. 35
426 Whirlpool Bathtubs.......................... 35

CHAPTER 5   WATER HEATERS................ 37
Section
501 General................................... 37
502 Installation................................ 37
503 Connections............................... 38
504 Safety Devices............................. 38
505 Insulation................................. 39

TABLE OF CONTENTS

CHAPTER 6 WATER SUPPLY AND DISTRIBUTION 41
Section
- 601 General 41
- 602 Water Required 41
- 603 Water Service 41
- 604 Design of Building Water Distribution System 42
- 605 Materials, Joints and Connections 44
- 606 Installation of the Building Water Distribution System 49
- 607 Hot Water Supply System 50
- 608 Protection of Potable Water Supply 51
- 609 Health Care Plumbing 57
- 610 Disinfection of Potable Water System 57
- 611 Drinking Water Treatment Units 58
- 612 Solar Systems 58
- 613 Temperature Control Devices and Valves 58

CHAPTER 7 SANITARY DRAINAGE 59
Section
- 701 General 59
- 702 Materials 59
- 703 Building Sewer 61
- 704 Drainage Piping Installation 61
- 705 Joints 62
- 706 Connections Between Drainage Piping and Fittings 64
- 707 Prohibited Joints and Connections 65
- 708 Cleanouts 65
- 709 Fixture Units 66
- 710 Drainage System Sizing 66
- 711 Offsets in Drainage Piping in Buildings of Five Stories or More 68
- 712 Sumps and Ejectors 68
- 713 Computerized Drainage Design 69
- 714 Backwater Valves 70
- 715 Vacuum Drainage Systems 70
- 716 Replacement of Underground Building Sewers and Building Drains by Pipe-Bursting Methods 70

CHAPTER 8 INDIRECT/SPECIAL WASTE 71
Section
- 801 General 71
- 802 Indirect Wastes 71
- 803 Special Wastes 72

CHAPTER 9 VENTS 73
Section
- 901 General 73
- 902 Materials 73
- 903 Vent Terminals 73
- 904 Outdoor Vent Extensions 73
- 905 Vent Connections and Grades 74
- 906 Vent Pipe Sizing 74
- 907 Vents for Stack Offsets 76
- 908 Relief Vents—Stacks of More Than 10 Branch Intervals 76
- 909 Fixture Vents 76
- 910 Individual Vent 76
- 911 Common Vent 77
- 912 Wet Venting 77
- 913 Waste Stack Vent 77
- 914 Circuit Venting 78
- 915 Combination Waste and Vent System 78
- 916 Island Fixture Venting 79
- 917 Single-Stack Vent System 79
- 918 Air Admittance Valves 80
- 919 Engineered Vent Systems 80
- 920 Computerized Vent Design 81

CHAPTER 10 TRAPS, INTERCEPTORS AND SEPARATORS 83
Section
- 1001 General 83
- 1002 Trap Requirements 83
- 1003 Interceptors and Separators 84
- 1004 Materials, Joints and Connections 86

CHAPTER 11 STORM DRAINAGE 87
Section
- 1101 General 87
- 1102 Materials 87
- 1103 Traps 88
- 1104 Conductors and Connections 88
- 1105 Roof Drains 88
- 1106 Size of Conductors, Leaders and Storm Drains 88
- 1107 Siphonic Roof Drainage Systems 94
- 1108 Secondary (Emergency) Roof Drains 94
- 1109 Combined Sanitary and Storm Public Sewer 95
- 1110 Controlled Flow Roof Drain Systems 95

1111	Subsoil Drains........................... 95
1112	Building Subdrains....................... 96
1113	Sumps and Pumping Systems............... 96

CHAPTER 12 SPECIAL PIPING AND STORAGE SYSTEMS............ 97
Section
1201	General................................. 97
1202	Medical Gases........................... 97
1203	Oxygen Systems......................... 97

CHAPTER 13 NONPOTABLE WATER SYSTEMS.............. 99
Section
1301	General................................. 99
1302	On-site Nonpotable Water Reuse Systems..... 101
1303	Nonpotable Rainwater Collection and Distribution Systems.................. 103
1304	Reclaimed Water Systems................. 105

CHAPTER 14 SUBSURFACE LANDSCAPE IRRIGATION SYSTEMS........ 107
Section
1401	General................................ 107
1402	System Design and Sizing................. 107
1403	Installation............................. 109

CHAPTER 15 REFERENCED STANDARDS..... 111

APPENDIX A PLUMBING PERMIT FEE SCHEDULE................ 127
Permit Issuance.................................... 127
Unit Fee Schedule................................. 127
Other Inspections and Fees........................ 127

APPENDIX B RATES OF RAINFALL FOR VARIOUS CITIES.......... 129

APPENDIX C STRUCTURAL SAFETY........ 131
Section
| C101 | Cutting, Notching and Boring in Wood Members...................... 131 |

APPENDIX D DEGREE DAY AND DESIGN TEMPERATURES....... 133

APPENDIX E SIZING OF WATER PIPING SYSTEM................ 139
Section
E101	General................................ 139
E102	Information Required.................... 139
E103	Selection of Pipe Size.................... 139
E201	Selection of Pipe Size.................... 156
E202	Determination of Pipe Volumes............ 156

INDEX.. 161

CHAPTER 1

SCOPE AND ADMINISTRATION

User note:

About this chapter: Chapter 1 establishes the limits of applicability of this code and describes how the code is to be applied and enforced. Chapter 1 is in two parts: Part 1—Scope and Application (Sections 101–102) and Part 2—Administration and Enforcement (Sections 103–110). Section 101 identifies which buildings and structures come under its purview and references other I-Codes as applicable. Standards and codes are scoped to the extent referenced (see Section 102.8).

This code is intended to be adopted as a legally enforceable document and it cannot be effective without adequate provisions for its administration and enforcement. The provisions of Chapter 1 establish the authority and duties of the code official appointed by the authority having jurisdiction and also establish the rights and privileges of the design professional, contractor and property owner.

PART 1—SCOPE AND APPLICATION

SECTION 101
GENERAL

[A] 101.1 Title. These regulations shall be known as the *Plumbing Code* of **[NAME OF JURISDICTION]** hereinafter referred to as "this code."

[A] 101.2 Scope. The provisions of this code shall apply to the erection, installation, alteration, repairs, relocation, replacement, addition to, use or maintenance of plumbing systems within this jurisdiction. This code shall regulate nonflammable medical gas, inhalation anesthetic, vacuum piping, nonmedical oxygen systems and sanitary and condensate vacuum collection systems. The installation of fuel gas distribution piping and equipment, fuel-gas-fired water heaters and water heater venting systems shall be regulated by the *International Fuel Gas Code*. Provisions in the appendices shall not apply unless specifically adopted.

> **Exception:** Detached one- and two-family dwellings and multiple single-family dwellings (townhouses) not more than three stories high with separate means of egress and their accessory structures shall comply with the *International Residential Code*.

[A] 101.3 Intent. The purpose of this code is to establish minimum standards to provide a reasonable level of safety, health, property protection and public welfare by regulating and controlling the design, construction, installation, quality of materials, location, operation and maintenance or use of plumbing equipment and systems.

[A] 101.4 Severability. If any section, subsection, sentence, clause or phrase of this code is for any reason held to be unconstitutional, such decision shall not affect the validity of the remaining portions of this code.

SECTION 102
APPLICABILITY

[A] 102.1 General. Where there is a conflict between a general requirement and a specific requirement, the specific requirement shall govern. Where, in any specific case, different sections of this code specify different materials, methods of construction or other requirements, the most restrictive shall govern.

[A] 102.2 Existing installations. Plumbing systems lawfully in existence at the time of the adoption of this code shall be permitted to have their use and maintenance continued if the use, maintenance or repair is in accordance with the original design and hazard to life, health or property is not created by such plumbing system.

> **[A] 102.2.1 Existing buildings.** Additions, alterations, renovations or repairs related to building or structural issues shall be regulated by the *International Existing Building Code*.

[A] 102.3 Maintenance. Plumbing systems, materials and appurtenances, both existing and new, and parts thereof, shall be maintained in proper operating condition in accordance with the original design in a safe and sanitary condition. Devices or safeguards required by this code shall be maintained in compliance with the edition of the code under which they were installed.

The owner or the owner's authorized agent shall be responsible for maintenance of plumbing systems. To determine compliance with this provision, the code official shall have the authority to require any plumbing system to be reinspected.

[A] 102.4 Additions, alterations or repairs. Additions, alterations, renovations or repairs to any plumbing system shall conform to that required for a new plumbing system without requiring the existing plumbing system to comply with all the requirements of this code. Additions, alterations or repairs shall not cause an existing system to become unsafe, insanitary or overloaded.

Minor additions, alterations, renovations and repairs to existing plumbing systems shall meet the provisions for new construction, unless such work is done in the same manner and arrangement as was in the existing system, is not hazardous and is *approved*.

[A] 102.5 Change in occupancy. It shall be unlawful to make any change in the *occupancy* of any structure that will subject the structure to any special provision of this code applicable to the new *occupancy* without approval of the code official. The code official shall certify that such structure meets the intent of the provisions of law governing building construction for the proposed new *occupancy* and that such change of *occupancy* does not result in any hazard to the public health, safety or welfare.

[A] 102.6 Historic buildings. The provisions of this code relating to the construction, alteration, repair, enlargement, restoration, relocation or moving of buildings or structures shall not be mandatory for existing buildings or structures identified and classified by the state or local jurisdiction as historic buildings where such buildings or structures are judged by the code official to be safe and in the public interest of health, safety and welfare regarding any proposed construction, alteration, repair, enlargement, restoration, relocation or moving of buildings.

[A] 102.7 Moved buildings. Except as determined by Section 102.2, plumbing systems that are a part of buildings or structures moved into or within the jurisdiction shall comply with the provisions of this code for new installations.

[A] 102.8 Referenced codes and standards. The codes and standards referenced in this code shall be those that are listed in Chapter 15 and such codes and standards shall be considered as part of the requirements of this code to the prescribed extent of each such reference and as further regulated in Sections 102.8.1 and 102.8.2.

[A] 102.8.1 Conflicts. Where conflicts occur between provisions of this code and the referenced standards, the provisions of this code shall apply.

[A] 102.8.2 Provisions in referenced codes and standards. Where the extent of the reference to a referenced code or standard includes subject matter that is within the scope of this code, the provisions of this code, as applicable, shall take precedence over the provisions in the referenced code or standard.

[A] 102.9 Requirements not covered by code. Any requirements necessary for the strength, stability or proper operation of an existing or proposed plumbing system, or for the public safety, health and general welfare, not specifically covered by this code shall be determined by the code official.

[A] 102.10 Other laws. The provisions of this code shall not be deemed to nullify any provisions of local, state or federal law.

[A] 102.11 Application of references. Reference to chapter section numbers, or to provisions not specifically identified by number, shall be construed to refer to such chapter, section or provision of this code.

PART 2—ADMINISTRATION AND ENFORCEMENT

SECTION 103
DEPARTMENT OF PLUMBING INSPECTION

[A] 103.1 General. The department of plumbing inspection is hereby created and the executive official in charge thereof shall be known as the code official.

[A] 103.2 Appointment. The code official shall be appointed by the chief appointing authority of the jurisdiction.

[A] 103.3 Deputies. In accordance with the prescribed procedures of this jurisdiction and with the concurrence of the appointing authority, the code official shall have the authority to appoint a deputy code official, other related technical officers, inspectors and other employees. Such employees shall have powers as delegated by the code official.

[A] 103.4 Liability. The code official, member of the board of appeals or employee charged with the enforcement of this code, while acting for the jurisdiction in good faith and without malice in the discharge of the duties required by this code or other pertinent law or ordinance, shall not thereby be rendered civilly or criminally liable personally, and is hereby relieved from all personal liability for any damage accruing to persons or property as a result of any act or by reason of an act or omission in the discharge of official duties.

[A] 103.4.1 Legal defense. Any suit or criminal complaint instituted against any officer or employee because of an act performed by that officer or employee in the lawful discharge of duties and under the provisions of this code shall be defended by the legal representative of the jurisdiction until the final termination of the proceedings. The code official or any subordinate shall not be liable for costs in any action, suit or proceeding that is instituted in pursuance of the provisions of this code.

SECTION 104
DUTIES AND POWERS OF THE CODE OFFICIAL

[A] 104.1 General. The code official is hereby authorized and directed to enforce the provisions of this code. The code official shall have the authority to render interpretations of this code and to adopt policies and procedures in order to clarify the application of its provisions. Such interpretations, policies and procedures shall be in compliance with the intent and purpose of this code. Such policies and procedures shall not have the effect of waiving requirements specifically provided for in this code.

[A] 104.2 Applications and permits. The code official shall receive applications, review construction documents and issue permits for the installation and alteration of plumbing systems, inspect the premises for which such permits have been issued, and enforce compliance with the provisions of this code.

[A] 104.3 Inspections. The code official shall make all the required inspections, or shall accept reports of inspection by *approved* agencies or individuals. Reports of such inspections shall be in writing and be certified by a responsible officer of such *approved agency* or by the responsible individual. The code official is authorized to engage such expert opinion as deemed necessary to report on unusual technical issues that arise, subject to the approval of the appointing authority.

[A] 104.4 Right of entry. Where it is necessary to make an inspection to enforce the provisions of this code, or where the code official has reasonable cause to believe that there exists in any building or on any premises any conditions or violations of this code that make the building or premises unsafe, insanitary, dangerous or hazardous, the code official shall have the authority to enter the building or premises at all reasonable times to inspect or to perform the duties imposed upon the code official by this code. If such building or premises is occupied, the code official shall present credentials to the occupant and request entry. If such building or premises is unoccupied, the code official shall first make a reasonable effort to locate the owner, the owner's authorized agent or other person having charge or control of the building or premises and request entry. If entry is refused, the code official shall have recourse to every remedy provided by law to secure entry.

Where the code official shall have first obtained a proper inspection warrant or other remedy provided by law to secure entry, the owner, owner's authorized agent, occupant or person having charge, care or control of any building or premises shall not fail or neglect, after proper request is made as herein provided, to promptly permit entry therein by the code official for the purpose of inspection and examination pursuant to this code.

[A] 104.5 Identification. The code official shall carry proper identification when inspecting structures or premises in the performance of duties under this code.

[A] 104.6 Notices and orders. The code official shall issue all necessary notices or orders to ensure compliance with this code.

[A] 104.7 Department records. The code official shall keep official records of applications received, permits and certificates issued, fees collected, reports of inspections, and notices and orders issued. Such records shall be retained in the official records for the period required for the retention of public records.

SECTION 105
APPROVAL

[A] 105.1 Modifications. Where there are practical difficulties involved in carrying out the provisions of this code, the code official shall have the authority to grant modifications for individual cases, upon application of the owner or owner's authorized agent, provided that the code official shall first find that special individual reason makes the strict letter of this code impractical and the modification conforms to the intent and purpose of this code and that such modification does not lessen health, life and fire safety requirements. The details of action granting modifications shall be recorded and entered in the files of the plumbing inspection department.

[A] 105.2 Alternative materials, design and methods of construction and equipment. The provisions of this code are not intended to prevent the installation of any material or to prohibit any design or method of construction not specifically prescribed by this code, provided that any such alternative has been *approved*. An alternative material or method of construction shall be *approved* where the code official finds that the proposed design is satisfactory and complies with the intent of the provisions of this code, and that the material, method or work offered is, for the purpose intended, not less than the equivalent of that prescribed in this code in quality, strength, effectiveness, fire resistance, durability and safety. Where the alternative material, design or method of construction is not *approved*, the code official shall respond in writing, stating the reasons why the alternative was not *approved*.

[A] 105.2.1 Research reports. Supporting data, where necessary to assist in the approval of materials or assemblies not specifically provided for in this code, shall consist of valid research reports from *approved* sources.

[A] 105.3 Required testing. Where there is insufficient evidence of compliance with the provisions of this code, or evidence that a material or method does not conform to the requirements of this code, or in order to substantiate claims for alternate materials or methods, the code official shall have the authority to require tests as evidence of compliance to be made at no expense to the jurisdiction.

[A] 105.3.1 Test methods. Test methods shall be as specified in this code or by other recognized test standards. In the absence of recognized and accepted test methods, the code official shall approve the testing procedures.

[A] 105.3.2 Testing agency. Tests shall be performed by an *approved agency*.

[A] 105.3.3 Test reports. Reports of tests shall be retained by the code official for the period required for retention of public records.

[A] 105.4 Approved materials and equipment. Materials, equipment and devices *approved* by the code official shall be constructed and installed in accordance with such approval.

[A] 105.4.1 Material and equipment reuse. Materials, equipment and devices shall not be reused unless such elements have been reconditioned, tested, placed in good and proper working condition and *approved*.

SECTION 106
PERMITS

[A] 106.1 Where required. Any owner, owner's authorized agent or contractor who desires to construct, enlarge, alter, repair, move, demolish or change the *occupancy* of a building or structure, or to erect, install, enlarge, alter, repair, remove, convert or replace any plumbing system, the installation of which is regulated by this code, or to cause any such work to be performed, shall first make application to the code official and obtain the required permit for the work.

[A] 106.1.1 Annual permit. Instead of an individual construction permit for each alteration to an already *approved* system or equipment or appliance installation, the code official is authorized to issue an annual permit upon application therefor to any person, firm or corporation regularly employing one or more qualified tradespersons in the building, structure or on the premises owned or operated by the applicant for the permit.

[A] 106.1.2 Annual permit records. The person to whom an annual permit is issued shall keep a detailed record of alterations made under such annual permit. The code official shall have access to such records at all times or such records shall be filed with the code official as designated.

[A] 106.2 Exempt work. The following work shall be exempt from the requirement for a permit:

1. The stopping of leaks in drains, water, soil, waste or vent pipe provided, however, that if any concealed trap, drainpipe, water, soil, waste or vent pipe becomes defective and it becomes necessary to remove and replace the same with new material, such work shall be considered as new work and a permit shall be obtained and inspection made as provided in this code.

2. The clearing of stoppages or the repairing of leaks in pipes, valves or fixtures, and the removal and reinstallation of water closets, provided that such repairs do not involve or require the replacement or rearrangement of valves, pipes or fixtures.

Exemption from the permit requirements of this code shall not be deemed to grant authorization for any work to be done in violation of the provisions of this code or any other laws or ordinances of this jurisdiction.

[A] 106.3 Application for permit. Each application for a permit, with the required fee, shall be filed with the code official on a form furnished for that purpose and shall contain a general description of the proposed work and its location. The application shall be signed by the owner or owner's authorized agent. The permit application shall indicate the proposed *occupancy* of all parts of the building and of that portion of the site or lot, if any, not covered by the building or structure and shall contain such other information required by the code official.

[A] 106.3.1 Construction documents. Construction documents, engineering calculations, diagrams and other such data shall be submitted in two or more sets with each application for a permit. The code official shall require construction documents, computations and specifications to be prepared and designed by a registered design professional where required by state law. Construction documents shall be drawn to scale and shall be of sufficient clarity to indicate the location, nature and extent of the work proposed and show in detail that the work conforms to the provisions of this code. Construction documents for buildings more than two stories in height shall indicate where penetrations will be made for pipes, fittings and components and shall indicate the materials and methods for maintaining required structural safety, fire-resistance rating and fireblocking.

Exception: The code official shall have the authority to waive the submission of construction documents, calculations or other data if the nature of the work applied for is such that reviewing of construction documents is not necessary to determine compliance with this code.

[A] 106.3.2 Preliminary inspection. Before a permit is issued, the code official shall be authorized to inspect and evaluate the systems, equipment, buildings, devices, premises and spaces or areas to be used.

[A] 106.3.3 Time limitation of application. An application for a permit for any proposed work shall be deemed to have been abandoned 180 days after the date of filing, unless such application has been pursued in good faith or a permit has been issued; except that the code official shall have the authority to grant one or more extensions of time for additional periods not exceeding 180 days each. The extension shall be requested in writing and justifiable cause demonstrated.

[A] 106.4 By whom application is made. Application for a permit shall be made by the person or agent to install all or part of any plumbing system. The applicant shall meet all qualifications established by statute, or by rules promulgated by this code, or by ordinance or by resolution. The full name and address of the applicant shall be stated in the application.

[A] 106.5 Permit issuance. The application, construction documents and other data filed by an applicant for permit shall be reviewed by the code official. If the code official finds that the proposed work conforms to the requirements of this code and all laws and ordinances applicable thereto, and that the fees specified in Section 106.6 have been paid, a permit shall be issued to the applicant.

[A] 106.5.1 Approved construction documents. When the code official issues the permit where construction documents are required, the construction documents shall be endorsed in writing and stamped "APPROVED." Such *approved* construction documents shall not be changed, modified or altered without authorization from the code official. Work shall be done in accordance with the *approved* construction documents.

The code official shall have the authority to issue a permit for the construction of a part of a plumbing system before the entire construction documents for the whole system have been submitted or *approved*, provided that adequate information and detailed statements have been filed complying with all pertinent requirements of this code. The holders of such permit shall proceed at their own risk without assurance that the permit for the entire plumbing system will be granted.

[A] 106.5.2 Validity. The issuance of a permit or approval of construction documents shall not be construed to be a permit for, or an approval of, any violation of any of the provisions of this code or any other ordinance of the jurisdiction. A permit presuming to give authority to violate or cancel the provisions of this code shall not be valid.

The issuance of a permit based on construction documents and other data shall not prevent the code official from thereafter requiring the correction of errors in said construction documents and other data or from preventing building operations being carried on thereunder where in violation of this code or of other ordinances of this jurisdiction.

[A] 106.5.3 Expiration. Every permit issued by the code official under the provisions of this code shall expire by limitation and become null and void if the work authorized by such permit is not commenced within 180 days from the date of such permit, or if the work authorized by such permit is suspended or abandoned at any time after the work is commenced for a period of 180 days. Before such work can be recommenced, a new permit shall be first obtained and the fee therefor shall be one-half the amount required for a new permit for such work, provided that changes have not been made and will not be made in the original construction documents for such work, and provided further that such suspension or abandonment has not exceeded 1 year.

[A] 106.5.4 Extensions. Any permittee holding an unexpired permit shall have the right to apply for an extension of the time within which the permittee will commence work under that permit when work is unable to be commenced within the time required by this section for good and satisfactory reasons. The code official shall extend the time for action by the permittee for a period not exceeding 180 days if there is reasonable cause. A permit shall not be extended more than once. The fee for an extension shall be one-half the amount required for a new permit for such work.

[A] 106.5.5 Suspension or revocation of permit. The code official shall have the authority to suspend or revoke a permit issued under the provisions of this code wherever the permit is issued in error or on the basis of incorrect, inaccurate or incomplete information, or in violation of any ordinance or regulation or any of the provisions of this code.

[A] 106.5.6 Retention of construction documents. One set of *approved* construction documents shall be retained by the code official for a period of not less than 180 days from date of completion of the permitted work, or as required by state or local laws.

One set of *approved* construction documents shall be returned to the applicant, and said set shall be kept on the site of the building or work at all times during which the work authorized thereby is in progress.

[A] 106.5.7 Previous approvals. This code shall not require changes in the construction documents, construction or designated *occupancy* of a structure for which a lawful permit has been heretofore issued or otherwise lawfully authorized, and the construction of which has been pursued in good faith within 180 days after the effective date of this code and has not been abandoned.

[A] 106.5.8 Posting of permit. The permit or a copy shall be kept on the site of the work until the completion of the project.

[A] 106.6 Fees. A permit shall not be issued until the fees prescribed in Section 106.6.2 have been paid, and an amendment to a permit shall not be released until the additional fee, if any, due to an increase of the plumbing systems, has been paid.

[A] 106.6.1 Work commencing before permit issuance. Any person who commences any work on a plumbing system before obtaining the necessary permits shall be subject to 100 percent of the usual permit fee in addition to the required permit fees.

[A] 106.6.2 Fee schedule. The fees for all plumbing work shall be as indicated in the following schedule:

[JURISDICTION TO INSERT APPROPRIATE SCHEDULE]

[A] 106.6.3 Fee refunds. The code official shall authorize the refunding of fees as follows:

1. The full amount of any fee paid hereunder that was erroneously paid or collected.

2. Not more than [SPECIFY PERCENTAGE] percent of the permit fee paid where work has been done under a permit issued in accordance with this code.

3. Not more than [SPECIFY PERCENTAGE] percent of the plan review fee paid where an application for a permit for which a plan review fee has been paid is withdrawn or canceled before any plan review effort has been expended.

The code official shall not authorize the refunding of any fee paid except upon written application filed by the original permittee not later than 180 days after the date of fee payment.

SECTION 107
INSPECTIONS AND TESTING

[A] 107.1 General. The code official is authorized to conduct such inspections as are deemed necessary to determine compliance with the provisions of this code. Construction or work for which a permit is required shall be subject to inspection by the code official, and such construction or work shall remain visible and able to be accessed for inspection purposes until *approved*. Approval as a result of an inspection shall not be construed to be an approval of a violation of the provisions of this code or of other ordinances of the jurisdiction. Inspections presuming to give authority to violate or cancel the provisions of this code or of other ordinances of the jurisdiction shall not be valid. It shall be the duty of the permit applicant to cause the work to remain visible and able to be accessed for inspection purposes. Neither the code official nor the jurisdiction shall be liable for expense entailed in the removal or replacement of any material required to allow inspection.

[A] 107.2 Required inspections and testing. The code official, upon notification from the permit holder or the permit holder's agent, shall make the following inspections and such other inspections as necessary, and shall either release that portion of the construction or shall notify the permit holder or an agent of any violations that must be corrected. The holder of the permit shall be responsible for the scheduling of such inspections.

1. Underground inspection shall be made after trenches or ditches are excavated and bedded, piping installed, and before any backfill is put in place.

2. Rough-in inspection shall be made after the roof, framing, fireblocking, firestopping, draftstopping and bracing is in place and all sanitary, storm and water distribution piping is roughed-in, and prior to the installation of wall or ceiling membranes.

3. Final inspection shall be made after the building is complete, all plumbing fixtures are in place and properly connected, and the structure is ready for occupancy.

[A] 107.2.1 Other inspections. In addition to the inspections specified in Section 107.2, the code official shall be authorized to make or require other inspections of any construction work to ascertain compliance with the provisions of this code and other laws that are enforced.

[A] 107.2.2 Inspection requests. It shall be the duty of the holder of the permit or their duly authorized agent to notify the code official when work is ready for inspection. It shall be the duty of the permit holder to provide *access* to and means for inspections of such work that are required by this code.

[A] 107.2.3 Approval required. Work shall not be done beyond the point indicated in each successive inspection without first obtaining the approval of the code official. The code official, upon notification, shall make the requested inspections and shall either indicate the portion of the construction that is satisfactory as completed, or notify the permit holder or his or her agent wherein the same fails to comply with this code. Any portions that do not comply shall be corrected and such portion shall not be covered or concealed until authorized by the code official.

[A] 107.2.4 Approved agencies. The code official is authorized to accept reports of *approved* inspection agencies, provided that such agencies satisfy the requirements as to qualifications and reliability.

[A] 107.2.5 Evaluation and follow-up inspection services. Prior to the approval of a closed, prefabricated plumbing system and the issuance of a plumbing permit, the code official shall require the submittal of an evaluation report on each prefabricated plumbing system indicating the complete details of the plumbing system, including a description of the system and its components, the basis on which the plumbing system is being evaluated, test results and similar information, and other data as necessary for the code official to determine conformance to this code.

[A] 107.2.5.1 Evaluation service. The code official shall designate the evaluation service of an *approved agency* as the evaluation agency, and review such agency's evaluation report for adequacy and conformance to this code.

[A] 107.2.5.2 Follow-up inspection. Except where *ready access* is provided to all plumbing systems, service equipment and accessories for complete inspection at the site without disassembly or dismantling, the code official shall conduct the frequency of in-plant inspections necessary to ensure conformance to the *approved* evaluation report or shall designate an independent, *approved* inspection agency to conduct such inspections. The inspection agency shall furnish the code official with the follow-up inspection manual and a report of inspections on request, and the plumbing system shall have an identifying label permanently affixed to the system indicating that factory inspections have been performed.

[A] 107.2.5.3 Test and inspection records. Required test and inspection records shall be available to the code official at all times during the fabrication of the plumbing system and the erection of the building, or such records as the code official designates shall be filed.

[A] 107.3 Special inspections. Special inspections of *alternative engineered design* plumbing systems shall be conducted in accordance with Sections 107.3.1 and 107.3.2.

[A] 107.3.1 Periodic inspection. The registered design professional or designated inspector shall periodically inspect and observe the *alternative engineered design* to determine that the installation is in accordance with the *approved* construction documents. Discrepancies shall be brought to the immediate attention of the plumbing contractor for correction. Records shall be kept of all inspections.

[A] 107.3.2 Written report. The registered design professional shall submit a final report in writing to the code official upon completion of the installation, certifying that the *alternative engineered design* conforms to the *approved* construction documents. A notice of approval for the plumbing system shall not be issued until a written certification has been submitted.

[A] 107.4 Testing. Plumbing work and systems shall be tested as required in Section 312 and in accordance with Sections 107.4.1 through 107.4.3. Tests shall be made by the permit holder and observed by the code official.

[A] 107.4.1 New, altered, extended or repaired systems. New plumbing systems and parts of existing systems that have been altered, extended or repaired shall be tested as prescribed herein to disclose leaks and defects, except that testing is not required in the following cases:

1. In any case that does not include addition to, replacement, alteration or relocation of any water supply, drainage or vent piping.
2. In any case where plumbing equipment is set up temporarily for exhibition purposes.

[A] 107.4.2 Equipment, material and labor for tests. Equipment, material and labor required for testing a plumbing system or part thereof shall be furnished by the permit holder.

[A] 107.4.3 Reinspection and testing. Where any work or installation does not pass any initial test or inspection, the necessary corrections shall be made to comply with this code. The work or installation shall then be resubmitted to the code official for inspection and testing.

[A] 107.5 Approval. After the prescribed tests and inspections indicate that the work complies in all respects with this code, a notice of approval shall be issued by the code official.

[A] 107.5.1 Revocation. The code official is authorized to, in writing, suspend or revoke a notice of approval issued under the provisions of this code wherever the notice is issued in error, or on the basis of incorrect information supplied, or where it is determined that the building or structure, premise or portion thereof is in violation of any ordinance or regulation or any of the provisions of this code.

[A] 107.6 Temporary connection. The code official shall have the authority to authorize the temporary connection of the building or system to the utility source for the purpose of testing plumbing systems or for use under a temporary certificate of occupancy.

[A] 107.7 Connection of service utilities. A person shall not make connections from a utility, source of energy, fuel, power, water system or *sewer* system to any building or system that is regulated by this code for which a permit is required until authorized by the code official.

SECTION 108
VIOLATIONS

[A] 108.1 Unlawful acts. It shall be unlawful for any person, firm or corporation to erect, construct, alter, repair, remove, demolish or utilize any plumbing system, or cause same to be done, in conflict with or in violation of any of the provisions of this code.

[A] 108.2 Notice of violation. The code official shall serve a notice of violation or order to the person responsible for the erection, installation, alteration, extension, repair, removal or demolition of plumbing work in violation of the provisions of this code, or in violation of a detail statement or the *approved* construction documents thereunder, or in violation of a permit or certificate issued under the provisions of this code. Such order shall direct the discontinuance of the illegal action or condition and the abatement of the violation.

[A] 108.3 Prosecution of violation. If the notice of violation is not complied with promptly, the code official shall request the legal counsel of the jurisdiction to institute the appropriate proceeding at law or in equity to restrain, correct or abate such violation, or to require the removal or termination of the unlawful occupancy of the structure in violation of the provisions of this code or of the order or direction made pursuant thereto.

[A] 108.4 Violation penalties. Any person who shall violate a provision of this code or shall fail to comply with any of the requirements thereof or who shall erect, install, alter or repair plumbing work in violation of the *approved* construction documents or directive of the code official, or of a permit or certificate issued under the provisions of this code, shall be guilty of a **[SPECIFY OFFENSE]**, punishable by a fine of not more than **[AMOUNT]** dollars or by imprisonment not exceeding **[NUMBER OF DAYS]**, or both such fine and imprisonment. Each day that a violation continues after due notice has been served shall be deemed a separate offense.

[A] 108.5 Stop work orders. Upon notice from the code official, work on any plumbing system that is being performed contrary to the provisions of this code or in a dangerous or unsafe manner shall immediately cease. Such notice shall be in writing and shall be given to the owner of the property, or to the owner's authorized agent, or to the person performing the work. The notice shall state the conditions under which work is authorized to resume. Where an emergency exists, the code official shall not be required to give a written notice prior to stopping the work. Any person who shall continue any work in or about the structure after having been served with a stop work order, except such work as that person is directed to perform to remove a violation or unsafe condition, shall be liable to a fine of not less than **[AMOUNT]** dollars or more than **[AMOUNT]** dollars.

[A] 108.6 Abatement of violation. The imposition of the penalties herein prescribed shall not preclude the legal officer of the jurisdiction from instituting appropriate action to prevent unlawful construction or to restrain, correct or abate a violation, or to prevent illegal occupancy of a building, structure or premises, or to stop an illegal act, conduct, business or utilization of the plumbing on or about any premises.

[A] 108.7 Unsafe plumbing. Any plumbing regulated by this code that is unsafe or that constitutes a fire or health hazard, insanitary condition, or is otherwise dangerous to human life is hereby declared unsafe. Any use of plumbing regulated by this code constituting a hazard to safety, health or public welfare by reason of inadequate maintenance, dilapidation, obsolescence, fire hazard, disaster, damage or abandonment is hereby declared an unsafe use. Any such unsafe equipment is hereby declared to be a public nuisance and shall be abated by repair, rehabilitation, demolition or removal.

[A] 108.7.1 Authority to condemn equipment. Where the code official determines that any plumbing, or portion thereof, regulated by this code has become hazardous to life, health or property or has become insanitary, the code official shall order in writing that such plumbing either be removed or restored to a safe or sanitary condition. A time limit for compliance with such order shall be specified in the written notice. A person shall not use or maintain defective plumbing after receiving such notice.

Where such plumbing is to be disconnected, written notice as prescribed in Section 108.2 shall be given. In cases of immediate danger to life or property, such disconnection shall be made immediately without such notice.

[A] 108.7.2 Authority to disconnect service utilities. The code official shall have the authority to authorize disconnection of utility service to the building, structure or system regulated by the technical codes in case of an emergency, where necessary, to eliminate an immediate danger to life or property. Where possible, the owner or the owner's authorized agent and occupant of the building, structure or service system shall be notified of the decision to disconnect utility service prior to taking such action. If not notified prior to disconnecting, the owner, the owner's authorized agent or occupant of the building, structure or service systems shall be notified in writing, as soon as practical thereafter.

[A] 108.7.3 Connection after order to disconnect. A person shall not make connections from any energy, fuel, power supply or water distribution system or supply energy, fuel or water to any equipment regulated by this code that has been disconnected or ordered to be disconnected by the code official or the use of which has been ordered to be discontinued by the code official until the code official authorizes the reconnection and use of such equipment.

Where any plumbing is maintained in violation of this code, and in violation of any notice issued pursuant to the provisions of this section, the code official shall institute any appropriate action to prevent, restrain, correct or abate the violation.

SECTION 109
MEANS OF APPEAL

[A] 109.1 Application for appeal. Any person shall have the right to appeal a decision of the code official to the board of appeals. An application for appeal shall be based on a claim that the true intent of this code or the rules legally adopted thereunder have been incorrectly interpreted, the provisions of this code do not fully apply, or an equally good or better form of construction is proposed. The application shall be filed on a form obtained from the code official within 20 days after the notice was served.

[A] 109.2 Membership of board. The board of appeals shall consist of five members appointed by the chief appointing authority as follows: one for 5 years, one for 4 years, one for 3 years, one for 2 years and one for 1 year. Thereafter, each new member shall serve for 5 years or until a successor has been appointed.

[A] 109.2.1 Qualifications. The board of appeals shall consist of five individuals, one from each of the following professions or disciplines:

1. Registered design professional who is a registered architect; or a builder or superintendent of building construction with not less than 10 years' experience, 5 years of which shall have been in responsible charge of work.

2. Registered design professional with structural engineering or architectural experience.

3. Registered design professional with mechanical and plumbing engineering experience; or a mechanical and plumbing contractor with not less than 10 years' experience, 5 years of which shall have been in responsible charge of work.

4. Registered design professional with electrical engineering experience; or an electrical contractor with not less than 10 years' experience, 5 years of which shall have been in responsible charge of work.

5. Registered design professional with fire protection engineering experience; or a fire protection contractor with not less than 10 years' experience, 5 years of which shall have been in responsible charge of work.

[A] 109.2.2 Alternate members. The chief appointing authority shall appoint two alternate members who shall be called by the board chairman to hear appeals during the absence or disqualification of a member. Alternate members shall possess the qualifications required for board membership, and shall be appointed for 5 years or until a successor has been appointed.

[A] 109.2.3 Chairman. The board shall annually select one of its members to serve as chairman.

[A] 109.2.4 Disqualification of member. A member shall not hear an appeal in which that member has any personal, professional or financial interest.

[A] 109.2.5 Secretary. The chief administrative officer shall designate a qualified clerk to serve as secretary to the board. The secretary shall file a detailed record of all proceedings in the office of the chief administrative officer.

[A] 109.2.6 Compensation of members. Compensation of members shall be determined by law.

[A] 109.3 Notice of meeting. The board shall meet upon notice from the chairman, within 10 days of the filing of an appeal or at stated periodic meetings.

[A] 109.4 Open hearing. Hearings before the board shall be open to the public. The appellant, the appellant's representative, the code official and any person whose interests are affected shall be given an opportunity to be heard.

[A] 109.4.1 Procedure. The board shall adopt and make available to the public through the secretary procedures under which a hearing will be conducted. The procedures shall not require compliance with strict rules of evidence, but shall mandate that only relevant information be received.

[A] 109.5 Postponed hearing. When five members are not present to hear an appeal, either the appellant or the appellant's representative shall have the right to request a postponement of the hearing.

[A] 109.6 Board decision. The board shall modify or reverse the decision of the code official by a concurring vote of three members.

[A] 109.6.1 Resolution. The decision of the board shall be by resolution. Certified copies shall be furnished to the appellant and to the code official.

[A] 109.6.2 Administration. The code official shall take immediate action in accordance with the decision of the board.

[A] 109.7 Court review. Any person, whether or not a previous party of the appeal, shall have the right to apply to the appropriate court for a writ of certiorari to correct errors of law. Application for review shall be made in the manner and time required by law following the filing of the decision in the office of the chief administrative officer.

SECTION 110
TEMPORARY EQUIPMENT, SYSTEMS AND USES

[A] 110.1 General. The code official is authorized to issue a permit for temporary equipment, systems and uses. Such permits shall be limited as to time of service, but shall not be permitted for more than 180 days. The code official is authorized to grant extensions for demonstrated cause.

[A] 110.2 Conformance. Temporary equipment, systems and uses shall conform to the structural strength, fire safety, means of egress, accessibility, light, ventilation and sanitary requirements of this code as necessary to ensure the public health, safety and general welfare.

[A] 110.3 Temporary utilities. The code official is authorized to give permission to temporarily supply utilities before an installation has been fully completed and the final certificate of completion has been issued. The part covered by the temporary certificate shall comply with the requirements specified for temporary lighting, heat or power in the code.

[A] 110.4 Termination of approval. The code official is authorized to terminate such permit for temporary equipment, systems or uses and to order the temporary equipment, systems or uses to be discontinued.

CHAPTER 2
DEFINITIONS

> **User note:**
> *About this chapter: Codes, by their very nature, are technical documents. Every word, term and punctuation mark can add to or change the meaning of a technical requirement. It is necessary to maintain a consensus on the specific meaning of each term contained in the code. Chapter 2 performs this function by stating clearly what specific terms mean for the purpose of the code.*

SECTION 201
GENERAL

201.1 Scope. Unless otherwise expressly stated, the following words and terms shall, for the purposes of this code, have the meanings shown in this chapter.

201.2 Interchangeability. Words stated in the present tense include the future; words stated in the masculine gender include the feminine and neuter; the singular number includes the plural and the plural the singular.

201.3 Terms defined in other codes. Where terms are not defined in this code and are defined in the *International Building Code*, *International Fire Code*, *International Fuel Gas Code* or the *International Mechanical Code*, such terms shall have the meanings ascribed to them as in those codes.

201.4 Terms not defined. Where terms are not defined through the methods authorized by this section, such terms shall have ordinarily accepted meanings such as the context implies.

SECTION 202
GENERAL DEFINITIONS

ACCEPTED ENGINEERING PRACTICE. That which conforms to accepted principles, tests or standards of nationally recognized technical or scientific authorities.

[M] ACCESS (TO). That which enables a fixture, appliance or equipment to be reached by ready *access* or by a means that first requires the removal or movement of a panel, door or similar obstruction (see "Ready *access*").

ACCESS COVER. A removable plate, usually secured by bolts or screws, to permit *access* to a pipe or pipe fitting for the purposes of inspection, repair or cleaning.

[BE] ACCESSIBLE. A site, building, facility or portion thereof that complies with Chapter 11 of the *International Building Code*.

ADAPTER FITTING. An *approved* connecting device that suitably and properly joins or adjusts pipes and fittings that do not otherwise fit together.

AIR ADMITTANCE VALVE. One-way valve designed to allow air to enter the plumbing drainage system when negative pressures develop in the piping system. The device shall close by gravity and seal the vent terminal at zero differential pressure (no-flow conditions) and under positive internal pressures. The purpose of an air admittance valve is to provide a method of allowing air to enter the plumbing drainage system without the use of a vent extended to open air and to prevent *sewer* gases from escaping into a building.

AIR BREAK (Drainage System). A piping arrangement in which a drain from a fixture, appliance or device discharges indirectly into another fixture, receptacle or interceptor at a point below the *flood level rim* and above the trap seal.

AIR GAP (Drainage System). The unobstructed vertical distance through the free atmosphere between the outlet of the waste pipe and the *flood level rim* of the receptacle into which the waste pipe is discharging.

AIR GAP (Water Distribution System). The unobstructed vertical distance through the free atmosphere between the lowest opening from any pipe or faucet supplying water to a tank, plumbing fixture or other device and the *flood level rim* of the receptacle.

ALTERNATE ON-SITE NONPOTABLE WATER. Nonpotable water from other than public utilities, on-site surface sources and subsurface natural freshwater sources. Examples of such water are graywater, on-site reclaimed water, collected rainwater, captured condensate and rejected water from reverse osmosis systems.

ALTERNATIVE ENGINEERED DESIGN. A plumbing system that performs in accordance with the intent of Chapters 3 through 14 and provides an equivalent level of performance for the protection of public health, safety and welfare. The system design is not specifically regulated by Chapters 3 through 14.

ANCHORS. See "Supports."

ANTISIPHON. A term applied to valves or mechanical devices that eliminate siphonage.

[A] APPROVED. Acceptable to the code official.

[A] APPROVED AGENCY. An established and recognized agency that is regularly engaged in conducting tests or furnishing inspection services, or furnishing product certification where such agency has been *approved* by the code official.

AREA DRAIN. A receptacle designed to collect surface or storm water from an open area.

BACKFLOW. Pressure created by any means in the water distribution system, which by being in excess of the pressure

DEFINITIONS

in the water supply mains causes a potential backflow condition.

   Backpressure, low head. A pressure less than or equal to 4.33 psi (29.88 kPa) or the pressure exerted by a 10-foot (3048 mm) column of water.

   Backsiphonage. The backflow of potentially contaminated water into the potable water system as a result of the pressure in the potable water system falling below atmospheric pressure of the plumbing fixtures, pools, tanks or vats connected to the potable water distribution piping.

   Water supply system. The flow of water or other liquids, mixtures or substances into the distribution pipes of a potable water supply from any source except the intended source.

BACKFLOW CONNECTION. Any arrangement whereby backflow is possible.

BACKFLOW, DRAINAGE. A reversal of flow in the drainage system.

BACKFLOW PREVENTER. A backflow prevention assembly, a backflow prevention device or other means or method to prevent backflow into the potable water supply.

BACKWATER VALVE. A device or valve installed in the *building drain* or *sewer* pipe where a *sewer* is subject to backflow, and that prevents drainage or waste from backing up into a lower level or fixtures and causing a flooding condition.

[BS] BASE FLOOD ELEVATION. A reference point, determined in accordance with the building code, based on the depth or peak elevation of flooding, including wave height, which has a 1 percent (100-year flood) or greater chance of occurring in any given year.

BATHROOM GROUP. A group of fixtures consisting of a water closet, lavatory, bathtub or shower, including or excluding a bidet, an *emergency floor drain* or both. Such fixtures are located together on the same floor level.

BRANCH. Any part of the piping system except a riser, main or *stack*.

BRANCH INTERVAL. A vertical measurement of distance, 8 feet (2438 mm) or more in *developed length*, between the connections of horizontal *branches* to a drainage *stack*. Measurements are taken down the *stack* from the highest horizontal *branch* connection.

BRANCH VENT. A vent connecting one or more individual vents with a vent *stack* or *stack* vent.

[A] BUILDING. Any structure utilized or intended for supporting or sheltering any occupancy.

BUILDING DRAIN. That part of the lowest piping of a drainage system that receives the discharge from soil, waste and other drainage pipes inside and that extends 30 inches (762 mm) in *developed length* of pipe beyond the exterior walls of the building and conveys the drainage to the *building sewer*.

   Combined. A *building drain* that conveys both sewage and storm water or other drainage.

   Sanitary. A *building drain* that conveys sewage only.

   Storm. A *building drain* that conveys storm water or other drainage, but not sewage.

BUILDING SEWER. That part of the drainage system that extends from the end of the *building drain* and conveys the discharge to a *public sewer*, *private sewer*, individual sewage disposal system or other point of disposal.

   Combined. A *building sewer* that conveys both sewage and storm water or other drainage.

   Sanitary. A *building sewer* that conveys sewage only.

   Storm. A *building sewer* that conveys storm water or other drainage, but not sewage.

BUILDING SUBDRAIN. That portion of a drainage system that does not drain by gravity into the *building sewer*.

BUILDING TRAP. A device, fitting or assembly of fittings installed in the *building drain* to prevent circulation of air between the drainage system of the building and the *building sewer*.

CIRCUIT VENT. A vent that connects to a horizontal drainage *branch* and vents two traps to not more than eight traps or trapped fixtures connected into a battery.

CIRCULATING HOT WATER SYSTEM. A specifically designed water distribution system where one or more pumps are operated in the service hot water piping to circulate heated water from the water-heating equipment to fixture supply and back to the water-heating equipment.

CISTERN. A small covered tank for storing water for a home or farm. Generally, this tank stores rainwater to be utilized for purposes other than in the potable water supply, and such tank is placed underground in most cases.

CLEANOUT. An access opening in the drainage system utilized for the removal of obstructions. Types of cleanouts include a removable plug or cap, and a removable fixture or fixture trap.

[A] CODE. These regulations, subsequent amendments thereto or any emergency rule or regulation that the administrative authority having jurisdiction has lawfully adopted.

[A] CODE OFFICIAL. The officer or other designated authority charged with the administration and enforcement of this code, or a duly authorized representative.

COLLECTION PIPE. Unpressurized pipe used within the collection system that drains on-site nonpotable water or rainwater to a storage tank by gravity.

COMBINATION FIXTURE. A fixture combining one sink and laundry tray or a two- or three-compartment sink or laundry tray in one unit.

COMBINATION WASTE AND VENT SYSTEM. A specially designed system of waste piping embodying the horizontal wet venting of one or more sinks, lavatories, drinking fountains or floor drains by means of a common waste and vent pipe adequately sized to provide free movement of air above the flow line of the drain.

COMBINED BUILDING DRAIN. See "*Building drain, combined.*"

COMBINED BUILDING SEWER. See "*Building sewer, combined.*"

COMMON VENT. A vent connecting at the junction of two *fixture drains* or to a fixture *branch* and serving as a vent for both fixtures.

CONCEALED FOULING SURFACE. Any surface of a plumbing fixture that is not readily visible and is not scoured or cleansed with each fixture operation.

CONDUCTOR. A pipe inside the building that conveys storm water from the roof to a storm or combined *building drain*.

[A] CONSTRUCTION DOCUMENT. All of the written, graphic and pictorial documents prepared or assembled for describing the design, location and physical characteristics of the elements of the project necessary for obtaining a building permit. The construction drawings shall be drawn to an appropriate scale.

CONTAMINATION. An impairment of the quality of the potable water that creates an actual hazard to the public health through poisoning or the spread of disease by sewage, industrial fluids or waste.

CRITICAL LEVEL (C-L). An elevation (height) reference point that determines the minimum height at which a backflow preventer or vacuum breaker is installed above the *flood level rim* of the fixture or receptor served by the device. The critical level is the elevation level below which there is a potential for backflow to occur. If the critical level marking is not indicated on the device, the bottom of the device shall constitute the critical level.

CROSS CONNECTION. Any physical connection or arrangement between two otherwise separate piping systems, one of which contains potable water and the other either water of unknown or questionable safety or steam, gas or chemical, whereby there exists the possibility for flow from one system to the other, with the direction of flow depending on the pressure differential between the two systems (see "Backflow").

DEMAND RECIRCULATION WATER SYSTEM. A water distribution system where one or more pumps prime the service hot water piping with heated water upon a demand for hot water.

DEPTH OF TRAP SEAL. The depth of liquid that would have to be removed from a full trap before air could pass through the trap.

[BS] DESIGN FLOOD ELEVATION. The elevation of the "design flood," including wave height, relative to the datum specified on the community's legally designated flood hazard map. In areas designated as Zone AO, the *design flood elevation* shall be the elevation of the highest existing grade of the building's perimeter plus the depth number (in feet) specified on the flood hazard map. In areas designated as Zone AO where a depth number is not specified on the map, the depth number shall be taken as being equal to 2 feet (610 mm).

DEVELOPED LENGTH. The length of a pipeline measured along the centerline of the pipe and fittings.

DISCHARGE PIPE. A pipe that conveys the discharge from plumbing fixtures or appliances.

DRAIN. Any pipe that carries wastewater or waterborne wastes in a building drainage system.

DRAINAGE FITTING. The type of fitting or fittings utilized in the drainage system. Drainage fittings are similar to cast-iron fittings, except that instead of having a bell and spigot, drainage fittings are recessed and tapped to eliminate ridges on the inside of the installed pipe.

DRAINAGE FIXTURE UNIT.

   Drainage (dfu). A measure of the probable discharge into the drainage system by various types of plumbing fixtures. The drainage fixture-unit value for a particular fixture depends on its volume rate of drainage discharge, on the time duration of a single drainage operation and on the average time between successive operations.

DRAINAGE SYSTEM. Piping within a *public* or *private* premise that conveys sewage, rainwater or other liquid waste to a point of disposal. A drainage system does not include the mains of a *public sewer* system or a private or public sewage treatment or disposal plant.

   Building gravity. A drainage system that drains by gravity into the *building sewer*.

   Sanitary. A drainage system that carries sewage and excludes storm, surface and ground water.

   Storm. A drainage system that carries rainwater, surface water, subsurface water and similar liquid waste.

DRINKING FOUNTAIN. A plumbing fixture that is connected to the potable water distribution system and the drainage system. The fixture allows the user to obtain a drink directly from a stream of flowing water without the use of any accessories.

EFFECTIVE OPENING. The minimum cross-sectional area at the point of water supply discharge, measured or expressed in terms of the diameter of a circle or, if the opening is not circular, the diameter of a circle of equivalent cross-sectional area. For faucets and similar fittings, the *effective opening* shall be measured at the smallest orifice in the fitting body or in the supply piping to the fitting.

EMERGENCY FLOOR DRAIN. A floor drain that does not receive the discharge of any drain or indirect waste pipe, and that protects against damage from accidental spills, fixture overflows and leakage.

ESSENTIALLY NONTOXIC TRANSFER FLUID. Fluids having a Gosselin rating of 1, including propylene glycol; mineral oil; polydimethylsiloxane; hydrochlorofluorocarbon, chlorofluorocarbon and carbon refrigerants; and FDA-approved boiler water additives for steam boilers.

ESSENTIALLY TOXIC TRANSFER FLUID. Soil, waste or graywater and fluids having a Gosselin rating of 2 or more, including ethylene glycol, hydrocarbon oils, ammonia refrigerants and hydrazine.

EXISTING INSTALLATION. Any plumbing system regulated by this code that was legally installed prior to the effective date of this code, or for which a permit to install has been issued.

FAUCET. A valve end of a water pipe through which water is drawn from or held within the pipe.

FILL VALVE. A water supply valve, opened or closed by means of a float or similar device, utilized to supply water to

DEFINITIONS

a tank. An antisiphon fill valve contains an antisiphon device in the form of an *approved air gap* or vacuum breaker that is an integral part of the fill valve unit and that is positioned on the discharge side of the water supply control valve.

FIXTURE. See "Plumbing fixture."

FIXTURE BRANCH. A drain serving two or more fixtures that discharges to another drain or to a *stack*.

FIXTURE DRAIN. The drain from the trap of a fixture to a junction with any other drain pipe.

FIXTURE FITTING.

  Supply fitting. A fitting that controls the volume, direction of flow or both of water and is either attached to or accessed from a fixture, or is used with an open or atmospheric discharge.

  Waste fitting. A combination of components that conveys the sanitary waste from the outlet of a fixture to the connection to the sanitary drainage system.

FIXTURE SUPPLY. The water supply pipe connecting a fixture to a *branch* water supply pipe or directly to a main water supply pipe.

[BS] FLOOD HAZARD AREA. The greater of the following two areas:

1. The area within a flood plain subject to a 1-percent or greater chance of flooding in any given year.
2. The area designated as a *flood hazard area* on a community's flood hazard map or as otherwise legally designated.

FLOOD LEVEL RIM. The edge of the receptacle from which water overflows.

FLOW CONTROL (Vented). A device installed upstream from the interceptor having an orifice that controls the rate of flow through the interceptor and an air intake (vent) downstream from the orifice that allows air to be drawn into the flow stream.

FLOW PRESSURE. The pressure in the water supply pipe near the faucet or water outlet while the faucet or water outlet is wide open and flowing.

FLUSH TANK. A tank designed with a fill valve and flush valve to flush the contents of the bowl or usable portion of the fixture.

FLUSHOMETER TANK. A device integrated within an air accumulator vessel that is designed to discharge a predetermined quantity of water to fixtures for flushing purposes.

FLUSHOMETER VALVE. A valve attached to a pressurized water supply pipe and designed so that when activated, the valve opens the line for direct flow into the fixture at a rate and quantity to operate the fixture properly, and then gradually closes to reseal fixture traps and avoid water hammer.

FULL-OPEN VALVE. A water control or shutoff component in the water supply system piping that, where adjusted for maximum flow, the flow path through the component's closure member is not a restriction in the component's through-flow area.

GRAYWATER. Waste discharged from lavatories, bathtubs, showers, clothes washers and laundry trays.

GREASE INTERCEPTOR.

  Fats, oils and greases (FOG) disposal system. A plumbing appurtenance that reduces nonpetroleum fats, oils and greases in effluent by separation or mass and volume reduction.

  Gravity. Plumbing appurtenances of not less than 500 gallons (1893 L) capacity that are installed in the sanitary drainage system to intercept free-floating fats, oils and grease from wastewater discharge. Separation is accomplished by gravity during a retention time of not less than 30 minutes.

  Hydromechanical. Plumbing appurtenances that are installed in the sanitary drainage system to intercept free-floating fats, oils and grease from wastewater discharge. Continuous separation is accomplished by air entrainment, buoyancy and interior baffling.

GREASE-LADEN WASTE. Effluent discharge that is produced from food processing, food preparation or other sources where grease, fats and oils enter automatic dishwater prerinse stations, sinks or other appurtenances.

GREASE REMOVAL DEVICE, AUTOMATIC (GRD). A plumbing appurtenance that is installed in the sanitary drainage system to intercept free-floating fats, oils and grease from wastewater discharge. Such a device operates on a time- or event-controlled basis and has the ability to remove free-floating fats, oils and grease automatically without intervention from the user except for maintenance.

GRIDDED WATER DISTRIBUTION SYSTEM. A water distribution system where every water distribution pipe is interconnected so as to provide two or more paths to each fixture supply pipe.

HANGERS. See "Supports."

HORIZONTAL BRANCH DRAIN. A drainage *branch* pipe extending laterally from a soil or waste *stack* or *building drain*, with or without vertical sections or *branches*, that receives the discharge from two or more *fixture drains* or *branches* and conducts the discharge to the soil or waste *stack* or to the *building drain*.

HORIZONTAL PIPE. Any pipe or fitting that makes an angle of less than 45 degrees (0.79 rad) with a horizontal plane.

HOT WATER. Water at a temperature greater than or equal to 110°F (43°C).

HOUSE TRAP. See "Building trap."

INDIRECT WASTE PIPE. A waste pipe that does not connect directly with the drainage system, but that discharges into the drainage system through an *air break* or *air gap* into a trap, fixture, receptor or interceptor.

INDIVIDUAL SEWAGE DISPOSAL SYSTEM. A system for disposal of domestic sewage by means of a septic tank, cesspool or mechanical treatment, designed for utilization apart from a public *sewer* to serve a single establishment or building.

INDIVIDUAL VENT. A pipe installed to vent a fixture trap and that connects with the vent system above the fixture served or terminates in the open air.

INDIVIDUAL WATER SUPPLY. A water supply that serves one or more families, and that is not an *approved* public water supply.

INTERCEPTOR. A device designed and installed to separate and retain for removal, by automatic or manual means, deleterious, hazardous or undesirable matter from normal wastes, while permitting normal sewage or wastes to discharge into the drainage system by gravity.

JOINT.

Expansion. A loop, return bend or return offset that provides for the expansion and contraction in a piping system and is utilized in tall buildings or where there is a rapid change of temperature, as in power plants, steam rooms and similar occupancies.

Flexible. Any joint between two pipes that permits one pipe to be deflected or moved without movement or deflection of the other pipe.

Mechanical. See "Mechanical joint."

Slip. A type of joint made by means of a washer or a special type of packing compound in which one pipe is slipped into the end of an adjacent pipe.

LEAD-FREE SOLDER AND FLUX. Containing not more than 0.2-percent lead.

LEADER. An exterior drainage pipe for conveying storm water from roof or gutter drains to an *approved* means of disposal.

MACERATING TOILET SYSTEM. An assembly consisting of a water closet and sump with a macerating pump that is designed to collect, grind and pump wastes from the water closet and up to two other fixtures connected to the sump.

MAIN. The principal pipe artery to which branches are connected.

MANIFOLD. See "Plumbing appurtenance."

[M] MECHANICAL JOINT. A connection between pipes, fittings, or pipes and fittings that is not screwed, caulked, threaded, soldered, solvent cemented, brazed, welded or heat fused. A joint in which compression is applied along the centerline of the pieces being joined. In some applications, the joint is part of a coupling, fitting or adapter.

MEDICAL GAS SYSTEM. The complete system to convey medical gases for direct patient application from central supply systems (bulk tanks, manifolds and medical air compressors), with pressure and operating controls, alarm warning systems, related components and piping networks extending to station outlet valves at patient use points.

MEDICAL VACUUM SYSTEM. A system consisting of central-vacuum-producing equipment with pressure and operating controls, shutoff valves, alarm-warning systems, gauges and a network of piping extending to and terminating with suitable station inlets at locations where patient suction may be required.

METER. A measuring device used to collect data and indicate water usage.

NONPOTABLE WATER. Water not safe for drinking, personal or culinary utilization.

NUISANCE. Public nuisance as known in common law or in equity jurisprudence; whatever is dangerous to human life or detrimental to health; whatever structure or premises is not sufficiently ventilated, sewered, drained, cleaned or lighted, with respect to its intended occupancy; and whatever renders the air, or human food, drink or water supply unwholesome.

[A] OCCUPANCY. The purpose for which a building or portion thereof is utilized or occupied.

OFFSET. A combination of *approved* bends that makes two changes in direction bringing one section of the pipe out of line but into a line parallel with the other section.

ON-SITE NONPOTABLE WATER REUSE SYSTEM. A water system for the collection, treatment, storage, distribution and reuse of nonpotable water generated on site, including but not limited to a graywater system. This definition does not include a rainwater harvesting system.

OPEN AIR. Outside the structure.

PLUMBING. The practice, materials and fixtures utilized in the installation, maintenance, extension and alteration of all piping, fixtures, plumbing appliances and plumbing appurtenances, within or adjacent to any structure, in connection with sanitary drainage or storm drainage facilities; venting systems; and public or private water supply systems.

PLUMBING APPLIANCE. Water or drain-connected devices intended to perform a special function. These devices have their operation or control dependent on one or more energized components, such as motors, controls or heating elements. Such devices are manually adjusted or controlled by the owner or operator, or are operated automatically through one or more of the following actions: a time cycle, a temperature range, a pressure range, a measured volume or weight.

PLUMBING APPURTENANCE. A manufactured device, prefabricated assembly or on-the-job assembly of component parts that is an adjunct to the basic piping system and plumbing fixtures. An appurtenance does not demand additional water supply and does not add any discharge load to a fixture or to the drainage system.

PLUMBING FIXTURE. A receptacle or device that is connected to a water supply system or discharges to a drainage system or both. Such receptacles or devices require a supply of water; or discharge liquid waste or liquid-borne solid waste; or require a supply of water and discharge waste to a drainage system.

PLUMBING SYSTEM. A system that includes the water distribution pipes; plumbing fixtures and traps; water-treating or water-using equipment; soil, waste and vent pipes; and *building drains*; in addition to their respective connections, devices and appurtenances within a structure or premises; and the water service, *building sewer* and building storm *sewer* serving such structure or premises.

POLLUTION. An impairment of the quality of the potable water to a degree that does not create a hazard to public health but that does adversely and unreasonably affect the aesthetic qualities of such potable water for domestic use.

POTABLE WATER. Water free from impurities present in amounts sufficient to cause disease or harmful physiological effects and conforming to the bacteriological and chemical quality requirements of the Public Health Service Drinking Water Standards or the regulations of the public health authority having jurisdiction.

[M] PRESS-CONNECT JOINT. A permanent mechanical joint incorporating an elastomeric seal or an elastomeric seal and corrosion-resistant grip ring. The joint is made with a pressing tool and jaw or ring approved by the fitting manufacturer.

PRIVATE. In the classification of plumbing fixtures, "*private*" applies to fixtures in residences and apartments, and to fixtures in nonpublic toilet rooms of hotels and motels and similar installations in buildings where the plumbing fixtures are intended for utilization by a family or an individual.

PUBLIC OR PUBLIC UTILIZATION. In the classification of plumbing fixtures, "*public*" applies to fixtures in general toilet rooms of schools, gymnasiums, hotels, airports, bus and railroad stations, public buildings, bars, public comfort stations, office buildings, stadiums, stores, restaurants and other installations where a number of fixtures are installed so that their utilization is similarly unrestricted.

PUBLIC SWIMMING POOL. A pool, other than a residential pool, that is intended to be used for swimming or bathing and is operated by an owner, lessee, operator, licensee or concessionaire, regardless of whether a fee is charged for use.

PUBLIC WATER MAIN. A water supply pipe for public utilization controlled by public authority.

QUICK-CLOSING VALVE. A valve or faucet that closes automatically when released manually or that is controlled by a mechanical means for fast-action closing.

RAINWATER. Water from natural precipitation.

[M] READY ACCESS. That which enables a fixture, appliance or equipment to be directly reached without requiring the removal or movement of any panel, door or similar obstruction and without the use of a portable ladder, step stool or similar device.

RECLAIMED WATER. Nonpotable water that has been derived from the treatment of wastewater by a facility or system licensed or permitted to produce water meeting the jurisdiction's water requirements for its intended uses. Also known as "recycled water."

REDUCED PRESSURE PRINCIPLE BACKFLOW PREVENTION ASSEMBLY. A backflow prevention device consisting of two independently acting check valves, internally force-loaded to a normally closed position and separated by an intermediate chamber (or zone) in which there is an automatic relief means of venting to the atmosphere, internally loaded to a normally open position between two tightly closing shutoff valves and with a means for testing for tightness of the checks and opening of the relief means.

[A] REGISTERED DESIGN PROFESSIONAL. An individual who is registered or licensed to practice their respective design profession, as defined by the statutory requirements of the professional registration laws of the state or jurisdiction in which the project is to be constructed.

RELIEF VALVE.

Pressure relief valve. A pressure-actuated valve held closed by a spring or other means and designed to relieve pressure automatically at the pressure at which such valve is set.

Temperature and pressure relief (T&P) valve. A combination relief valve designed to function as both a temperature relief and a pressure relief valve.

Temperature relief valve. A temperature-actuated valve designed to discharge automatically at the temperature at which such valve is set.

RELIEF VENT. A vent whose primary function is to provide circulation of air between drainage and vent systems.

RIM. An unobstructed open edge of a fixture.

RISER. See "Water pipe, riser."

ROOF DRAIN. A drain installed to receive water collecting on the surface of a roof and to discharge such water into a leader or a conductor.

ROUGH-IN. Parts of the plumbing system that are installed prior to the installation of fixtures. This includes drainage, water supply, vent piping and the necessary fixture supports and any fixtures that are built into the structure.

SELF-CLOSING FAUCET. A faucet containing a valve that automatically closes upon deactivation of the opening means.

SEPARATOR. See "Interceptor."

SEWAGE. Any liquid waste containing animal or vegetable matter in suspension or solution, including liquids containing chemicals in solution.

SEWAGE EJECTOR. A device for lifting sewage by entraining the sewage in a high-velocity jet of steam, air or water.

SEWER.

Building sewer. See "Building sewer."

Public sewer. That part of the drainage system of pipes, installed and maintained by a city, township, county, public utility company or other public entity, and located on public property, in the street or in an approved dedicated easement of public or community use.

Sanitary sewer. A *sewer* that carries sewage and excludes storm, surface and ground water.

Storm sewer. A *sewer* that conveys rainwater, surface water, subsurface water and similar liquid wastes.

SLOPE. The fall (pitch) of a line of pipe in reference to a horizontal plane. In drainage, the slope is expressed as the fall in units vertical per units horizontal (percent) for a length of pipe.

SOIL PIPE. A pipe that conveys sewage containing fecal matter to the *building drain* or *building sewer*.

SPILLPROOF VACUUM BREAKER. An assembly consisting of one check valve force-loaded closed and an air-inlet vent valve force-loaded open to atmosphere, positioned downstream of the check valve, and located between and including two tightly closing shutoff valves and a test cock.

STACK. A general term for any vertical line of soil, waste, vent or inside conductor piping that extends through not fewer than one story with or without offsets.

STACK VENT. The extension of a soil or waste *stack* above the highest horizontal drain connected to the *stack*.

STACK VENTING. A method of venting a fixture or fixtures through the soil or waste *stack*.

STORM DRAIN. See "Drainage system, storm."

STORM WATER. Natural precipitation, including snowmelt, that has contacted a surface at or below grade.

[A] STRUCTURE. That which is built or constructed.

SUBSOIL DRAIN. A drain that collects subsurface water or seepage water and conveys such water to a place of disposal.

SUMP. A tank or pit that receives sewage or liquid waste, located below the normal grade of the gravity system and that must be emptied by mechanical means.

SUMP PUMP. An automatic water pump powered by an electric motor for the removal of drainage, except raw sewage, from a sump, pit or low point.

SUMP VENT. A vent from pneumatic sewage ejectors, or similar equipment, that terminates separately to the open air.

SUPPORTS. Devices for supporting and securing pipe, fixtures and equipment.

SWIMMING POOL. A permanent or temporary structure that is intended to be used for swimming, bathing or wading and that is designed and manufactured or built to be connected to a circulation system. A swimming pool can be open to the public regardless of whether a fee is charged for its use or can be accessory to a residential setting where the pool is available only to the household and guests of the household.

TEMPERED WATER. Water having a temperature range between 85°F (29°C) and 110°F (43°C).

THIRD-PARTY CERTIFICATION AGENCY. An *approved* agency operating a product or material certification system that incorporates initial product testing, assessment and surveillance of a manufacturer's quality control system.

THIRD-PARTY CERTIFIED. Certification obtained by the manufacturer indicating that the function and performance characteristics of a product or material have been determined by testing and ongoing surveillance by an *approved third-party certification agency*. Assertion of certification is in the form of identification in accordance with the requirements of the *third-party certification agency*.

TOILET FACILITY. A room or space that contains not less than one water closet and one lavatory.

TRAP. A fitting or device that provides a liquid seal to prevent the emission of *sewer* gases without materially affecting the flow of sewage or wastewater through the trap.

TRAP SEAL. The vertical distance between the weir and the top of the dip of the trap.

UNSTABLE GROUND. Earth that does not provide a uniform bearing for the barrel of the *sewer* pipe between the joints at the bottom of the pipe trench.

VACUUM. Any pressure less than that exerted by the atmosphere.

VACUUM BREAKER. A type of backflow preventer installed on openings subject to normal atmospheric pressure that prevents backflow by admitting atmospheric pressure through ports to the discharge side of the device.

VENT PIPE. See "Vent system."

VENT STACK. A vertical vent pipe installed primarily for the purpose of providing circulation of air to and from any part of the drainage system.

VENT SYSTEM. A pipe or pipes installed to provide a flow of air to or from a drainage system, or to provide a circulation of air within such system to protect trap seals from siphonage and backpressure.

VERTICAL PIPE. Any pipe or fitting that makes an angle of 45 degrees (0.79 rad) or more with the horizontal.

WALL-HUNG WATER CLOSET. A wall-mounted water closet installed in such a way that the fixture does not touch the floor.

WASTE. The discharge from any fixture, appliance, area or appurtenance that does not contain fecal matter.

WASTE PIPE. A pipe that conveys only waste.

WASTE RECEPTOR. A floor sink, standpipe, hub drain or floor drain that receives the discharge of one or more indirect waste pipes.

WATER COOLER. A drinking fountain that incorporates a means of reducing the temperature of the water supplied to it from the potable water distribution system.

WATER DISPENSER. A plumbing fixture that is manually controlled by the user for the purpose of dispensing potable drinking water into a receptacle such as a cup, glass or bottle. Such fixture is connected to the potable water distribution system of the premises. This definition includes a freestanding apparatus for the same purpose that is not connected to the potable water distribution system and that is supplied with potable water from a container, bottle or reservoir.

WATER-HAMMER ARRESTOR. A device utilized to absorb the pressure surge (water hammer) that occurs when water flow is suddenly stopped in a water supply system.

[M] WATER HEATER. Any heating appliance or equipment that heats potable water and supplies such water to the potable hot water distribution system.

WATER MAIN. A water supply pipe or system of pipes, installed and maintained by a city, township, county, public utility company or other public entity, on public property, in the street or in an approved dedicated easement of public or community use.

WATER OUTLET. A discharge opening through which water is supplied to a fixture, into the atmosphere (except into an open tank that is part of the water supply system), to a boiler or heating system, or to any devices or equipment that

DEFINITIONS

require water to operate but are not part of the plumbing system.

WATER PIPE.

    Riser. A water supply pipe that extends one full story or more to convey water to *branches* or to a group of fixtures.

    Water distribution pipe. A pipe within the structure or on the premises that conveys water from the water service pipe, or from the meter when the meter is at the structure, to the points of utilization.

    Water service pipe. The pipe from the water main or other source of potable water supply, or from the meter when the meter is at the public right of way, to the water distribution system of the building served.

WATER SUPPLY SYSTEM. The water service pipe, water distribution pipes, and the necessary connecting pipes, fittings, control valves and all appurtenances in or adjacent to the structure or premises.

WELL.

    Bored. A well constructed by boring a hole in the ground with an auger and installing a casing.

    Drilled. A well constructed by making a hole in the ground with a drilling machine of any type and installing a casing and screen.

    Driven. A well constructed by driving a pipe in the ground. The drive pipe is usually fitted with a well point and screen.

    Dug. A well constructed by excavating a large-diameter shaft and installing a casing.

WHIRLPOOL BATHTUB. A plumbing appliance consisting of a bathtub fixture that is equipped and fitted with a circulating piping system designed to accept, circulate and discharge bathtub water upon each use.

YOKE VENT. A pipe connecting upward from a soil or waste *stack* to a vent *stack* for the purpose of preventing pressure changes in the *stacks*.

CHAPTER 3

GENERAL REGULATIONS

User note:

About this chapter: Chapter 3 covers general regulations for plumbing installations. As many of these requirements would need to be repeated in Chapters 3 through 14, placing such requirements in only one location eliminates code development coordination issues associated with the same requirement in multiple locations. These general requirements can be superseded by more specific requirements for certain applications in Chapters 3 through 14.

SECTION 301
GENERAL

301.1 Scope. The provisions of this chapter shall govern the general regulations regarding the installation of plumbing not specific to other chapters.

301.2 System installation. Plumbing shall be installed with due regard to preservation of the strength of structural members and prevention of damage to walls and other surfaces through fixture usage.

301.3 Connections to drainage system. Plumbing fixtures, drains, appurtenances and appliances used to receive or discharge liquid waste or sewage shall be directly connected to the sanitary drainage system of the building or premises, in accordance with the requirements of this code. This section shall not be construed to prevent indirect waste systems required by Chapter 8.

> **Exception:** Bathtubs, showers, lavatories, clothes washers and laundry trays shall not be required to discharge to the sanitary drainage system where such fixtures discharge to an *approved* system in accordance with Chapters 13 and 14.

301.4 Connections to water supply. Every plumbing fixture, device or appliance requiring or using water for its proper operation shall be directly or indirectly connected to the water supply system in accordance with the provisions of this code.

301.5 Pipe, tube and fitting sizes. Unless otherwise indicated, the pipe, tube and fitting sizes specified in this code are expressed in nominal or standard sizes as designated in the referenced material standards.

301.6 Prohibited locations. Plumbing systems shall not be located in an elevator shaft or in an elevator equipment room.

> **Exception:** Floor drains, sumps and sump pumps shall be permitted at the base of the shaft, provided that they are indirectly connected to the plumbing system and comply with Section 1003.4.

301.7 Conflicts. In instances where conflicts occur between this code and the manufacturer's installation instructions, the more restrictive provisions shall apply.

SECTION 302
EXCLUSION OF MATERIALS DETRIMENTAL TO THE SEWER SYSTEM

302.1 Detrimental or dangerous materials. Ashes, cinders or rags; flammable, poisonous or explosive liquids or gases; oil, grease or any other insoluble material capable of obstructing, damaging or overloading the building drainage or *sewer* system, or capable of interfering with the normal operation of the sewage treatment processes, shall not be deposited, by any means, into such systems.

302.2 Industrial wastes. Waste products from manufacturing or industrial operations shall not be introduced into the public *sewer* until it has been determined by the code official or other authority having jurisdiction that the introduction thereof will not damage the public *sewer* system or interfere with the functioning of the sewage treatment plant.

SECTION 303
MATERIALS

303.1 Identification. Each length of pipe and each pipe fitting, trap, fixture, material and device utilized in a plumbing system shall bear the identification of the manufacturer and any markings required by the applicable referenced standards.

303.2 Installation of materials. Materials used shall be installed in strict accordance with the standards under which the materials are accepted and *approved*. In the absence of such installation procedures, the manufacturer's instructions shall be followed. Where the requirements of referenced standards or manufacturer's installation instructions do not conform to minimum provisions of this code, the provisions of this code shall apply.

303.3 Plastic pipe, fittings and components. Plastic pipe, fittings and components shall be third-party certified as conforming to NSF 14.

303.4 Third-party certification. Plumbing products and materials required by the code to be in compliance with a referenced standard shall be listed by a *third-party certification agency* as complying with the referenced standards. Products and materials shall be identified in accordance with Section 303.1.

303.5 Cast-iron soil pipe, fittings and components. Cast-iron soil pipes and fittings, and the couplings used to join these products together, shall be third-party listed and labeled. Third-party certifiers or inspectors shall comply with the minimum inspection requirements of Annex A or Annex A1 of the ASTM and CISPI product standards indicated in the code for such products.

SECTION 304
RODENTPROOFING

304.1 General. Plumbing systems shall be designed and installed in accordance with Sections 304.2 through 304.4 to prevent rodents from entering structures.

304.2 Strainer plates. Strainer plates on drain inlets shall be designed and installed so that all openings are not greater than $^1/_2$ inch (12.7 mm) in least dimension.

304.3 Meter boxes. Meter boxes shall be constructed in such a manner that rodents are prevented from entering a structure by way of the water service pipes connecting the meter box and the structure.

304.4 Openings for pipes. In or on structures where openings have been made in walls, floors or ceilings for the passage of pipes, the annular space between the pipe and the sides of the opening shall be sealed with caulking materials or closed with gasketing systems compatible with the piping materials and locations.

SECTION 305
PROTECTION OF PIPES AND PLUMBING SYSTEM COMPONENTS

305.1 Protection against contact. Metallic piping, except for cast iron, ductile iron and galvanized steel, shall not be placed in direct contact with steel framing members, concrete or cinder walls and floors or other masonry. Metallic piping shall not be placed in direct contact with corrosive soil. Where sheathing is used to prevent direct contact, the sheathing shall have a thickness of not less than 0.008 inch (8 mil) (0.203 mm) and the sheathing shall be made of plastic. Where sheathing protects piping that penetrates concrete or masonry walls or floors, the sheathing shall be installed in a manner that allows movement of the piping within the sheathing.

305.2 Stress and strain. Piping in a plumbing system shall be installed so as to prevent strains and stresses that exceed the structural strength of the pipe. Where necessary, provisions shall be made to protect piping from damage resulting from expansion, contraction and structural settlement.

305.3 Pipes through foundation walls. Any pipe that passes through a foundation wall shall be provided with a relieving arch, or a pipe sleeve pipe shall be built into the foundation wall. The sleeve shall be two pipe sizes greater than the pipe passing through the wall.

305.4 Freezing. Water, soil and waste pipes shall not be installed outside of a building, in attics or crawl spaces, concealed in outside walls, or in any other place subjected to freezing temperatures unless adequate provision is made to protect such pipes from freezing by insulation or heat or both. Exterior water supply system piping shall be installed not less than 6 inches (152 mm) below the frost line and not less than 12 inches (305 mm) below grade.

305.4.1 Sewer depth. *Building sewers* that connect to private sewage disposal systems shall be installed not less than [NUMBER] inches (mm) below finished grade at the point of septic tank connection. *Building sewers* shall be installed not less than [NUMBER] inches (mm) below grade.

305.5 Waterproofing of openings. Joints at the roof and around vent pipes shall be made watertight by the use of lead, copper, galvanized steel, aluminum, plastic or other *approved* flashings or flashing material. Exterior wall openings shall be made watertight.

305.6 Protection against physical damage. In concealed locations where piping, other than cast-iron or galvanized steel, is installed through holes or notches in studs, joists, rafters or similar members less than $1^1/_4$ inches (32 mm) from the nearest edge of the member, the pipe shall be protected by steel shield plates. Such shield plates shall have a thickness of not less than 0.0575 inch (1.463 mm) (No. 16 gage). Such plates shall cover the area of the pipe where the member is notched or bored, and shall extend not less than 2 inches (51 mm) above sole plates and below top plates.

305.7 Protection of components of plumbing system. Components of a plumbing system installed along alleyways, driveways, parking garages or other locations exposed to damage shall be recessed into the wall or otherwise protected in an *approved* manner.

SECTION 306
TRENCHING, EXCAVATION AND BACKFILL

306.1 Support of piping. Buried piping shall be supported throughout its entire length.

306.2 Trenching and bedding. Where trenches are excavated such that the bottom of the trench forms the bed for the pipe, solid and continuous load-bearing support shall be provided between joints. Bell holes, hub holes and coupling holes shall be provided at points where the pipe is joined. Such pipe shall not be supported on blocks to grade. In instances where the materials manufacturer's installation instructions are more restrictive than those prescribed by the code, the material shall be installed in accordance with the more restrictive requirement.

306.2.1 Overexcavation. Where trenches are excavated below the installation level of the pipe such that the bottom of the trench does not form the bed for the pipe, the trench shall be backfilled to the installation level of the bottom of the pipe with sand or fine gravel placed in layers not greater than 6 inches (152 mm) in depth and such backfill shall be compacted after each placement.

306.2.2 Rock removal. Where rock is encountered in trenching, the rock shall be removed to not less than 3 inches (76 mm) below the installation level of the bottom of the pipe, and the trench shall be backfilled to the installation level of the bottom of the pipe with sand tamped in place so as to provide uniform load-bearing support for the pipe between joints. The pipe, including the joints, shall not rest on rock at any point.

306.2.3 Soft load-bearing materials. If soft materials of poor load-bearing quality are found at the bottom of the trench, stabilization shall be achieved by overexcavating not less than two pipe diameters and backfilling to the installation level of the bottom of the pipe with fine gravel, crushed stone or a concrete foundation. The concrete foundation shall be bedded with sand tamped into place so as to provide uniform load-bearing support for the pipe between joints.

306.3 Backfilling. Backfill shall be free from discarded construction material and debris. Loose earth free from rocks, broken concrete and frozen chunks shall be placed in the trench in 6-inch (152 mm) layers and tamped in place until the crown of the pipe is covered by 12 inches (305 mm) of tamped earth. The backfill under and beside the pipe shall be compacted for pipe support. Backfill shall be brought up evenly on both sides of the pipe so that the pipe remains aligned. In instances where the manufacturer's instructions for materials are more restrictive than those prescribed by the code, the material shall be installed in accordance with the more restrictive requirement.

306.4 Tunneling. Where pipe is to be installed by tunneling, jacking or a combination of both, the pipe shall be protected from damage during installation and from subsequent uneven loading. Where earth tunnels are used, adequate supporting structures shall be provided to prevent future settling or caving.

SECTION 307
STRUCTURAL SAFETY

307.1 General. In the process of installing or repairing any part of a plumbing and drainage installation, the finished floors, walls, ceilings, tile work or any other part of the building or premises that must be changed or replaced shall be left in a safe structural condition in accordance with the requirements of the *International Building Code*.

307.2 Cutting, notching or bored holes. A framing member shall not be cut, notched or bored in excess of limitations specified in the *International Building Code*.

307.3 Penetrations of floor/ceiling assemblies and fire-resistance-rated assemblies. Penetrations of floor/ceiling assemblies and assemblies required to have a fire-resistance rating shall be protected in accordance with the *International Building Code*.

[BS] 307.4 Alterations to trusses. Truss members and components shall not be cut, drilled, notched, spliced or otherwise altered in any way without written concurrence and approval of a registered design professional. Alterations resulting in the addition of loads to any member (such as HVAC equipment and water heaters) shall not be permitted without verification that the truss is capable of supporting such additional loading.

307.5 Protection of footings. Trenching installed parallel to footings and walls shall not extend into the bearing plane of a footing or wall. The upper boundary of the bearing plane is a line that extends downward, at an angle of 45 degrees (0.79 rad) from horizontal, from the outside bottom edge of the footing or wall.

307.6 Piping materials exposed within plenums. Piping materials exposed within plenums shall comply with the provisions of the *International Mechanical Code*.

SECTION 308
PIPING SUPPORT

308.1 General. Plumbing piping shall be supported in accordance with this section.

308.2 Piping seismic supports. Where earthquake loads are applicable in accordance with the building code, plumbing piping supports shall be designed and installed for the seismic forces in accordance with the *International Building Code*.

308.3 Materials. Hangers, anchors and supports shall support the piping and the contents of the piping. Hangers and strapping material shall be of *approved* material that will not promote galvanic action.

308.4 Structural attachment. Hangers and anchors shall be attached to the building construction in an *approved* manner.

308.5 Interval of support. Pipe shall be supported in accordance with Table 308.5.

> **Exception:** The interval of support for piping systems designed to provide for expansion/contraction shall conform to the engineered design in accordance with Section 316.1.

308.6 Sway bracing. Where horizontal pipes 4 inches (102 mm) and larger convey drainage or waste, and where a pipe fitting in that piping changes the flow direction greater than 45 degrees (0.79 rad), rigid bracing or other rigid support arrangements shall be installed to resist movement of the upstream pipe in the direction of pipe flow. A change of flow direction into a vertical pipe shall not require the upstream pipe to be braced.

308.7 Anchorage. Anchorage shall be provided to restrain drainage piping from axial movement.

> **308.7.1 Location.** For pipe sizes greater than 4 inches (102 mm), restraints shall be provided for drain pipes at all changes in direction and at all changes in diameter greater than two pipe sizes. Braces, blocks, rodding and other suitable methods as specified by the coupling manufacturer shall be utilized.

308.8 Expansion joint fittings. Expansion joint fittings shall be used only where necessary to provide for expansion and contraction of the pipes. Expansion joint fittings shall be of the typical material suitable for use with the type of piping in which such fittings are installed.

308.9 Parallel water distribution systems. Piping bundles for manifold systems shall be supported in accordance with Table 308.5. Support at changes in direction shall be in accordance with the manufacturer's instructions. Where hot water piping is bundled with cold or hot water piping, each hot water pipe shall be insulated.

308.10 Thermal expansion tanks. A thermal expansion tank shall be supported in accordance with the manufacturer's

GENERAL REGULATIONS

instructions. Thermal expansion tanks shall not be supported by the piping that connects to such tanks.

**TABLE 308.5
HANGER SPACING**

PIPING MATERIAL	MAXIMUM HORIZONTAL SPACING (feet)	MAXIMUM VERTICAL SPACING (feet)
Acrylonitrile butadiene styrene (ABS) pipe	4	10[b]
Aluminum tubing	10	15
Brass pipe	10	10
Cast-iron pipe	5[a]	15
Chlorinated polyvinyl chloride (CPVC) pipe and tubing, 1 inch and smaller	3	10[b]
Chlorinated polyvinyl chloride (CPVC) pipe and tubing, 1¼ inches and larger	4	10[b]
Copper or copper-alloy pipe	12	10
Copper or copper-alloy tubing, 1¼-inch diameter and smaller	6	10
Copper or copper-alloy tubing, 1½-inch diameter and larger	10	10
Cross-linked polyethylene (PEX) pipe 1 inch and smaller	2.67 (32 inches)	10[b]
Cross-linked polyethylene (PEX) pipe 1¼ inch and larger	4	10[b]
Cross-linked polyethylene/ aluminum/cross-linked polyethylene (PEX-AL-PEX) pipe	2.67 (32 inches)	4
Lead pipe	Continuous	4
Polyethylene/aluminum/ polyethylene (PE-AL-PE) pipe	2.67 (32 inches)	4
Polyethylene of raised temperature (PE-RT) pipe 1 inch and smaller	2.67 (32 inches)	10[b]
Polyethylene of raised temperature (PE-RT) pipe 1¼ inch and larger	4	10[b]
Polypropylene (PP) pipe or tubing 1 inch and smaller	2.67 (32 inches)	10[b]
Polypropylene (PP) pipe or tubing, 1¼ inches and larger	4	10[b]
Polyvinyl chloride (PVC) pipe	4	10[b]
Stainless steel drainage systems	10	10[b]
Steel pipe	12	15

For SI: 1 inch = 25.4 mm, 1 foot = 304.8 mm.

a. The maximum horizontal spacing of cast-iron pipe hangers shall be increased to 10 feet where 10-foot lengths of pipe are installed.

b. For sizes 2 inches and smaller, a guide shall be installed midway between required vertical supports. Such guides shall prevent pipe movement in a direction perpendicular to the axis of the pipe.

SECTION 309
FLOOD HAZARD RESISTANCE

309.1 General. Plumbing systems and equipment in structures erected in *flood hazard areas* shall be constructed in accordance with the requirements of this section and the *International Building Code*.

[BS] 309.2 Flood hazard. For structures located in *flood hazard areas*, the following systems and equipment shall be located and installed as required by Section 1612 of the *International Building Code*.

1. Water service pipes.
2. Pump seals in individual water supply systems where the pump is located below the *design flood elevation*.
3. Covers on potable water wells shall be sealed, except where the top of the casing well or pipe sleeve is elevated to not less than 1 foot (305 mm) above the *design flood elevation*.
4. Sanitary drainage piping.
5. Storm drainage piping.
6. Manhole covers shall be sealed, except where elevated to or above the *design flood elevation*.
7. Other plumbing fixtures, faucets, fixture fittings, piping systems and equipment.
8. Water heaters.
9. Vents and vent systems.

Exception: The systems listed in this section are permitted to be located below the elevation required by Section 1612 of the *International Building Code* for utilities and attendant equipment, provided that the systems are designed and installed to prevent water from entering or accumulating within their components and the systems are constructed to resist hydrostatic and hydrodynamic loads and stresses, including the effects of buoyancy, during the occurrence of flooding up to such elevation.

[BS] 309.3 Coastal high-hazard areas and coastal A zones. Structures located in coastal high-hazard areas and coastal A zones shall meet the requirements of Section 309.2. The plumbing systems, pipes and fixtures shall not be mounted on or penetrate through walls intended to break away under flood loads.

SECTION 310
WASHROOM AND TOILET ROOM REQUIREMENTS

310.1 Light and ventilation. Washrooms and toilet rooms shall be illuminated and ventilated in accordance with the *International Building Code* and *International Mechanical Code*.

310.2 Location of fixtures and compartments. The location of plumbing fixtures and the requirements for compartments and partitions shall be in accordance with Section 405.3.

310.3 Interior finish. Interior finish surfaces of toilet rooms shall comply with the *International Building Code*.

SECTION 311
TOILET FACILITIES FOR WORKERS

311.1 General. Toilet facilities shall be provided for construction workers and such facilities shall be maintained in a sanitary condition. Construction worker toilet facilities of the nonsewer type shall conform to PSAI Z4.3.

SECTION 312
TESTS AND INSPECTIONS

312.1 Required tests. The permit holder shall make the applicable tests prescribed in Sections 312.2 through 312.10 to determine compliance with the provisions of this code. The permit holder shall give reasonable advance notice to the code official when the plumbing work is ready for tests. The equipment, material, power and labor necessary for the inspection and test shall be furnished by the permit holder and he or she shall be responsible for determining that the work will withstand the test pressure prescribed in the following tests. Plumbing system piping shall be tested with either water or, for piping systems other than plastic, by air. After the plumbing fixtures have been set and their traps filled with water, the entire drainage system shall be submitted to final tests. The code official shall require the removal of any cleanouts if necessary to ascertain whether the pressure has reached all parts of the system.

312.1.1 Test gauges. Gauges used for testing shall be as follows:

1. Tests requiring a pressure of 10 pounds per square inch (psi) (69 kPa) or less shall utilize a testing gauge having increments of 0.10 psi (0.69 kPa) or less.

2. Tests requiring a pressure of greater than 10 psi (69 kPa) but less than or equal to 100 psi (689 kPa) shall utilize a testing gauge having increments of 1 psi (6.9 kPa) or less.

3. Tests requiring a pressure of greater than 100 psi (689 kPa) shall utilize a testing gauge having increments of 2 psi (14 kPa) or less.

312.2 Drainage and vent water test. A water test shall be applied to the drainage system either in its entirety or in sections. If applied to the entire system, all openings in the piping shall be tightly closed, except the highest opening, and the system shall be filled with water to the point of overflow. If the system is tested in sections, each opening shall be tightly plugged except the highest openings of the section under test, and each section shall be filled with water, but sections shall not be tested with less than a 10-foot (3048 mm) head of water. In testing successive sections, not less than the upper 10 feet (3048 mm) of the next preceding section shall be tested so that no joint or pipe in the building, except the uppermost 10 feet (3048 mm) of the system, shall have been submitted to a test of less than a 10-foot (3048 mm) head of water. This pressure shall be held for not less than 15 minutes. The system shall then be tight at all points.

312.3 Drainage and vent air test. Plastic piping shall not be tested using air. An air test shall be made by forcing air into the system until there is a uniform gauge pressure of 5 psi (34.5 kPa) or sufficient to balance a 10-inch (254 mm) column of mercury. This pressure shall be held for a test period of not less than 15 minutes. Any adjustments to the test pressure required because of changes in ambient temperatures or the seating of gaskets shall be made prior to the beginning of the test period.

312.4 Drainage and vent final test. The final test of the completed drainage and vent systems shall be visual and in sufficient detail to determine compliance with the provisions of this code. Where a smoke test is utilized, it shall be made by filling all traps with water and then introducing into the entire system a pungent, thick smoke produced by one or more smoke machines. When the smoke appears at *stack* openings on the roof, the *stack* openings shall be closed and a pressure equivalent to a 1-inch water column (248.8 Pa) shall be held for a test period of not less than 15 minutes.

312.5 Water supply system test. Upon completion of a section of or the entire water supply system, the system, or portion completed, shall be tested and proved tight under a water pressure not less than the working pressure of the system; or, for piping systems other than plastic, by an air test of not less than 50 psi (344 kPa). This pressure shall be held for not less than 15 minutes. The water utilized for tests shall be obtained from a potable source of supply. The required tests shall be performed in accordance with this section and Section 107.

312.6 Gravity sewer test. Gravity *sewer* tests shall consist of plugging the end of the *building sewer* at the point of connection with the public sewer, filling the *building sewer* with water, testing with not less than a 10-foot (3048 mm) head of water and maintaining such pressure for 15 minutes.

312.7 Forced sewer test. Forced *sewer* tests shall consist of plugging the end of the *building sewer* at the point of connection with the public sewer and applying a pressure of 5 psi (34.5 kPa) greater than the pump rating, and maintaining such pressure for 15 minutes.

312.8 Storm drainage system test. *Storm drain* systems within a building shall be tested by water or air in accordance with Section 312.2 or 312.3.

312.9 Shower liner test. Where shower floors and receptors are made watertight by the application of materials required by Section 421.5.2, the completed liner installation shall be tested. The pipe from the shower drain shall be plugged watertight for the test. The floor and receptor area shall be filled with potable water to a depth of not less than 2 inch (51 mm) measured at the threshold. Where a threshold of 2 inches (51 mm) high or greater does not exist, a temporary threshold shall be constructed to retain the test water in the lined floor or receptor area to a level not less than 2 inches (51 mm) deep measured at the threshold. The water shall be retained for a test period of not less than 15 minutes, and there shall not be evidence of leakage.

312.10 Inspection and testing of backflow prevention assemblies. Inspection and testing shall comply with Sections 312.10.1 and 312.10.2.

312.10.1 Inspections. Annual inspections shall be made of all backflow prevention assemblies and *air gaps* to determine whether the assemblies are operable and air gaps exist.

312.10.2 Testing. Reduced pressure principle, double check, pressure vacuum breaker, reduced pressure detector fire protection, double check detector fire protection, and spill-resistant vacuum breaker backflow preventer assemblies and hose connection backflow preventers shall be tested at the time of installation, immediately after repairs or relocation and at least annually. The testing procedure shall be performed in accordance with one of the following standards: ASSE 5013, ASSE 5015, ASSE 5020, ASSE 5047, ASSE 5048, ASSE 5052, ASSE 5056, CSA B64.10 or CSA B64.10.1.

SECTION 313
EQUIPMENT EFFICIENCIES

313.1 General. Equipment efficiencies shall be in accordance with the *International Energy Conservation Code*.

SECTION 314
CONDENSATE DISPOSAL

[M] 314.1 Fuel-burning appliances. Liquid combustion by-products of condensing appliances shall be collected and discharged to an *approved* plumbing fixture or disposal area in accordance with the manufacturer's instructions. Condensate piping shall be of *approved* corrosion-resistant material and shall not be smaller than the drain connection on the appliance. Such piping shall maintain a horizontal slope in the direction of discharge of not less than one-eighth unit vertical in 12 units horizontal (1-percent slope).

[M] 314.2 Evaporators and cooling coils. Condensate drain systems shall be provided for equipment and appliances containing evaporators or cooling coils. Condensate drain systems shall be designed, constructed and installed in accordance with Sections 314.2.1 through 314.2.5.

[M] 314.2.1 Condensate disposal. Condensate from all cooling coils and evaporators shall be conveyed from the drain pan outlet to an *approved* place of disposal. Such piping shall maintain a horizontal slope in the direction of discharge of not less than one-eighth unit vertical in 12 units horizontal (1-percent slope). Condensate shall not discharge into a street, alley or other areas so as to cause a nuisance.

[M] 314.2.2 Drain pipe materials and sizes. Components of the condensate disposal system shall be cast iron, galvanized steel, copper and copper alloy, cross-linked polyethylene, polyethylene, ABS, CPVC, PVC or polypropylene pipe or tubing. Components shall be selected for the pressure and temperature rating of the installation. Joints and connections shall be made in accordance with the applicable provisions of Chapter 7 relative to the material type. Condensate waste and drain line size shall be not less than $^3/_4$-inch (19.1 mm) internal diameter and shall not decrease in size from the drain pan connection to the place of condensate disposal. Where the drain pipes from more than one unit are manifolded together for condensate drainage, the pipe or tubing shall be sized in accordance with Table 314.2.2.

**[M] TABLE 314.2.2
CONDENSATE DRAIN SIZING**

EQUIPMENT CAPACITY	MINIMUM CONDENSATE PIPE DIAMETER (inch)
Up to 20 tons of refrigeration	$^3/_4$ inch
Over 20 tons to 40 tons of refrigeration	1 inch
Over 40 tons to 90 tons of refrigeration	$1^1/_4$ inch
Over 90 tons to 125 tons of refrigeration	$1^1/_2$ inch
Over 125 tons to 250 tons of refrigeration	2 inch

For SI: 1 inch = 25.4 mm, 1 ton of capacity = 3.517 kW.

[M] 314.2.3 Auxiliary and secondary drain systems. In addition to the requirements of Section 314.2.1, where damage to any building components could occur as a result of overflow from the equipment primary condensate removal system, one of the following auxiliary protection methods shall be provided for each cooling coil or fuel-fired appliance that produces condensate:

1. An auxiliary drain pan with a separate drain shall be provided under the coils on which condensation will occur. The auxiliary pan drain shall discharge to a conspicuous point of disposal to alert occupants in the event of a stoppage of the primary drain. The pan shall have a depth of not less than $1^1/_2$ inches (38 mm), shall be not less than 3 inches (76 mm) larger than the unit or the coil dimensions in width and length and shall be constructed of corrosion-resistant material. Galvanized sheet metal pans shall have a thickness of not less than 0.0236-inch (0.6010 mm) (No. 24 gage) galvanized sheet metal. Nonmetallic pans shall have a thickness of not less than 0.0625 inch (1.6 mm).

2. A separate overflow drain line shall be connected to the drain pan provided with the equipment. Such overflow drain shall discharge to a conspicuous point of disposal to alert occupants in the event of a stoppage of the primary drain. The overflow drain line shall connect to the drain pan at a higher level than the primary drain connection.

3. An auxiliary drain pan without a separate drain line shall be provided under the coils on which condensate will occur. Such pan shall be equipped with a water-level detection device conforming to UL 508 that will shut off the equipment served prior to overflow of the pan. The auxiliary drain pan shall be constructed in accordance with Item 1 of this section.

4. A water-level detection device conforming to UL 508 shall be provided that will shut off the equipment served in the event that the primary drain is blocked. The device shall be installed in the primary drain line, the overflow drain line or in the equipment-supplied drain pan, located at a point higher

than the primary drain line connection and below the overflow rim of such pan.

> **Exception:** Fuel-fired appliances that automatically shut down operation in the event of a stoppage in the condensate drainage system.

[M] 314.2.3.1 Water-level monitoring devices. On down-flow units and all other coils that do not have a secondary drain or provisions to install a secondary or auxiliary drain pan, a water-level monitoring device shall be installed inside the primary drain pan. This device shall shut off the equipment served in the event that the primary drain becomes restricted. Devices installed in the drain line shall not be permitted.

[M] 314.2.3.2 Appliance, equipment and insulation in pans. Where appliances, equipment or insulation are subject to water damage when auxiliary drain pans fill such portions of the appliances, equipment and insulation shall be installed above the *flood level rim* of the pan. Supports located inside of the pan to support the appliance or equipment shall be water resistant and *approved*.

[M] 314.2.4 Traps. Condensate drains shall be trapped as required by the equipment or appliance manufacturer.

[M] 314.2.4.1 Ductless mini-split system traps. Ductless mini-split equipment that produces condensation shall be provided with an in-line check valve located in the drain line or a trap.

[M] 314.2.5 Drain line maintenance. Condensate drain lines shall be configured to permit the clearing of blockages and performance of maintenance without requiring the drain line to be cut.

SECTION 315
PENETRATIONS

315.1 Sealing of annular spaces. The annular space between the outside of a pipe and the inside of a pipe sleeve or between the outside of a pipe and an opening in a building envelope wall, floor, or ceiling assembly penetrated by a pipe shall be sealed in an *approved* manner with caulking material, foam sealant or closed with a gasketing system. The caulking material, foam sealant or gasketing system shall be designed for the conditions at the penetration location and shall be compatible with the pipe, sleeve and building materials in contact with the sealing materials. Annular spaces created by pipes penetrating fire-resistance-rated assemblies or membranes of such assemblies shall be sealed or closed in accordance with Section 714 of the *International Building Code*.

SECTION 316
ALTERNATIVE ENGINEERED DESIGN

316.1 Alternative engineered design. The design, documentation, inspection, testing and approval of an *alternative engineered design* plumbing system shall comply with Sections 316.1.1 through 316.1.6.

316.1.1 Design criteria. An *alternative engineered design* shall conform to the intent of the provisions of this code and shall provide an equivalent level of quality, strength, effectiveness, fire resistance, durability and safety. Material, equipment or components shall be designed and installed in accordance with the manufacturer's instructions.

316.1.2 Submittal. The registered design professional shall indicate on the permit application that the plumbing system is an *alternative engineered design*. The permit and permanent permit records shall indicate that an *alternative engineered design* was part of the *approved* installation.

316.1.3 Technical data. The registered design professional shall submit sufficient technical data to substantiate the proposed *alternative engineered design* and to prove that the performance meets the intent of this code.

316.1.4 Construction documents. The registered design professional shall submit to the code official two complete sets of signed and sealed construction documents for the *alternative engineering design*. The construction documents shall include floor plans and a riser diagram of the work. Where appropriate, the construction documents shall indicate the direction of flow, all pipe sizes, grade of horizontal piping, loading and location of fixtures and appliances.

316.1.5 Design approval. Where the code official determines that the *alternative engineered design* conforms to the intent of this code, the plumbing system shall be *approved*. If the *alternative engineered design* is not *approved*, the code official shall notify the registered design professional in writing, stating the reasons thereof.

316.1.6 Inspection and testing. The *alternative engineered design* shall be tested and inspected in accordance with the requirements of Sections 107 and 312.

CHAPTER 4

FIXTURES, FAUCETS AND FIXTURE FITTINGS

User note:

About this chapter: Plumbing fixtures are required to be installed for nearly every building as toilet facilities (water closets and lavatories) are needed by the occupants of a building. Additional fixtures for washing, bathing and culinary purposes are also necessary where occupants dwell in buildings. Chapter 4 specifies the minimum number and type of plumbing fixtures for buildings based on the description of use of the building. Because fixture design and quality are paramount to ensure that plumbing fixtures operate properly, this chapter also specifies numerous product and material standards for plumbing fixtures.

SECTION 401
GENERAL

401.1 Scope. This chapter shall govern the materials, design and installation of plumbing fixtures, faucets and fixture fittings in accordance with the type of *occupancy*, and shall provide for the minimum number of fixtures for various types of occupancies.

401.2 Prohibited fixtures and connections. Water closets having a concealed trap seal or an unventilated space or having walls that are not thoroughly washed at each discharge in accordance with ASME A112.19.2/CSA B45.1 shall be prohibited. Any water closet that permits siphonage of the contents of the bowl back into the tank shall be prohibited. Trough urinals shall be prohibited.

401.3 Water conservation. The maximum water flow rates and flush volume for plumbing fixtures and fixture fittings shall comply with Section 604.4.

SECTION 402
FIXTURE MATERIALS

402.1 Quality of fixtures. Plumbing fixtures shall be constructed of *approved* materials, with smooth, impervious surfaces, free from defects and concealed fouling surfaces, and shall conform to standards cited in this code. Porcelain enameled surfaces on plumbing fixtures shall be acid resistant.

402.2 Materials for specialty fixtures. Materials for specialty fixtures not otherwise covered in this code shall be of stainless steel, soapstone, chemical stoneware or plastic, or shall be lined with lead, copper-base alloy, nickel-copper alloy, corrosion-resistant steel or other material especially suited to the application for which the fixture is intended.

402.3 Sheet copper. Sheet copper for general applications shall conform to ASTM B152 and shall not weigh less than 12 ounces per square foot (3.7 kg/m^2).

402.4 Sheet lead. Sheet lead for pans shall not weigh less than 4 pounds per square foot (19.5 kg/m^2) and shall be coated with an asphalt paint or other *approved* coating.

SECTION 403
MINIMUM PLUMBING FACILITIES

403.1 Minimum number of fixtures. Plumbing fixtures shall be provided in the minimum number as shown in Table 403.1, based on the actual use of the building or space. Uses not shown in Table 403.1 shall be considered individually by the code official. The number of occupants shall be determined by the *International Building Code*.

TABLE 403.1
MINIMUM NUMBER OF REQUIRED PLUMBING FIXTURES[a]
(See Sections 403.1.1 and 403.2)

NO.	CLASSIFICATION	DESCRIPTION	WATER CLOSETS (URINALS: SEE SECTION 424.2)		LAVATORIES		BATHTUBS/ SHOWERS	DRINKING FOUNTAIN (SEE SECTION 410)	OTHER
			MALE	FEMALE	MALE	FEMALE			
1	Assembly	Theaters and other buildings for the performing arts and motion pictures[d]	1 per 125	1 per 65	1 per 200		—	1 per 500	1 service sink
		Nightclubs, bars, taverns, dance halls and buildings for similar purposes[d]	1 per 40	1 per 40	1 per 75		—	1 per 500	1 service sink
		Restaurants, banquet halls and food courts[d]	1 per 75	1 per 75	1 per 200		—	1 per 500	1 service sink

(continued)

FIXTURES, FAUCETS AND FIXTURE FITTINGS

TABLE 403.1 —continued
MINIMUM NUMBER OF REQUIRED PLUMBING FIXTURES[a]
(See Sections 403.1.1 and 403.2)

NO.	CLASSIFICATION	DESCRIPTION	WATER CLOSETS (URINALS: SEE SECTION 424.2)		LAVATORIES		BATHTUBS/ SHOWERS	DRINKING FOUNTAIN (SEE SECTION 410)	OTHER
			MALE	FEMALE	MALE	FEMALE			
1 (cont.)	Assembly	Casino gaming areas	1 per 100 for the first 400 and 1 per 250 for the remainder exceeding 400	1 per 50 for the first 400 and 1 per 150 for the remainder exceeding 400	1 per 250 for the first 750 and 1 per 500 for the remainder exceeding 750		—	1 per 1,000	1 service sink
		Auditoriums without permanent seating, art galleries, exhibition halls, museums, lecture halls, libraries, arcades and gymnasiums[d]	1 per 125	1 per 65	1 per 200		—	1 per 500	1 service sink
		Passenger terminals and transportation facilities[d]	1 per 500	1 per 500	1 per 750		—	1 per 1,000	1 service sink
		Places of worship and other religious services[d]	1 per 150	1 per 75	1 per 200		—	1 per 1,000	1 service sink
		Coliseums, arenas, skating rinks, pools and tennis courts for indoor sporting events and activities	1 per 75 for the first 1,500 and 1 per 120 for the remainder exceeding 1,500	1 per 40 for the first 1,520 and 1 per 60 for the remainder exceeding 1,520	1 per 200	1 per 150	—	1 per 1,000	1 service sink
		Stadiums, amusement parks, bleachers and grandstands for outdoor sporting events and activities[f]	1 per 75 for the first 1,500 and 1 per 120 for the remainder exceeding 1,500	1 per 40 for the first 1,520 and 1 per 60 for the remainder exceeding 1,520	1 per 200	1 per 150	—	1 per 1,000	1 service sink
2	Business	Buildings for the transaction of business, professional services, other services involving merchandise, office buildings, banks, ambulatory care, light industrial and similar uses	1 per 25 for the first 50 and 1 per 50 for the remainder exceeding 50		1 per 40 for the first 80 and 1 per 80 for the remainder exceeding 80		—	1 per 100	1 service sink[e]
3	Educational	Educational facilities	1 per 50		1 per 50		—	1 per 100	1 service sink
4	Factory and industrial	Structures in which occupants are engaged in work fabricating, assembly or processing of products or materials	1 per 100		1 per 100		—	1 per 400	1 service sink
5	Institutional	Custodial care facilities	1 per 10		1 per 10		1 per 8	1 per 100	1 service sink
		Medical care recipients in hospitals and nursing homes	1 per room[c]		1 per room[c]		1 per 15	1 per 100	1 service sink per floor
		Employees in hospitals and nursing homes[b]	1 per 25		1 per 35		—	1 per 100	—
		Visitors in hospitals and nursing homes	1 per 75		1 per 100		—	1 per 500	—
		Prisons[b]	1 per cell		1 per cell		1 per 15	1 per 100	1 service sink
		Reformitories, detention centers, and correctional centers[b]	1 per 15		1 per 15		1 per 15	1 per 100	1 service sink
		Employees in reformitories, detention centers and correctional centers[b]	1 per 25		1 per 35		—	1 per 100	—
		Adult day care and child day care	1 per 15		1 per 15		1	1 per 100	1 service sink

(continued)

FIXTURES, FAUCETS AND FIXTURE FITTINGS

TABLE 403.1 —continued
MINIMUM NUMBER OF REQUIRED PLUMBING FIXTURES[a]
(See Sections 403.1.1 and 403.2)

NO.	CLASSIFICATION	DESCRIPTION	WATER CLOSETS (URINALS: SEE SECTION 424.2)		LAVATORIES		BATHTUBS/ SHOWERS	DRINKING FOUNTAIN (SEE SECTION 410)	OTHER
			MALE	FEMALE	MALE	FEMALE			
6	Mercantile	Retail stores, service stations, shops, salesrooms, markets and shopping centers	1 per 500		1 per 750		—	1 per 1,000	1 service sink[e]
7	Residential	Hotels, motels, boarding houses (transient)	1 per sleeping unit		1 per sleeping unit		1 per sleeping unit	—	1 service sink
		Dormitories, fraternities, sororities and boarding houses (not transient)	1 per 10		1 per 10		1 per 8	1 per 100	1 service sink
		Apartment house	1 per dwelling unit		1 per dwelling unit		1 per dwelling unit	—	1 kitchen sink per dwelling unit; 1 automatic clothes washer connection per 20 dwelling units
		Congregate living facilities with 16 or fewer persons	1 per 10		1 per 10		1 per 8	1 per 100	1 service sink
		One- and two-family dwellings and lodging houses with five or fewer guestrooms	1 per dwelling unit		1 per dwelling unit		1 per dwelling unit	—	1 kitchen sink per dwelling unit; 1 automatic clothes washer connection per dwelling unit
8	Storage	Structures for the storage of goods, warehouses, storehouse and freight depots. Low and Moderate Hazard.	1 per 100		1 per 100		—	1 per 1,000	1 service sink

a. The fixtures shown are based on one fixture being the minimum required for the number of persons indicated or any fraction of the number of persons indicated. The number of occupants shall be determined by the *International Building Code*.
b. Toilet facilities for employees shall be separate from facilities for inmates or care recipients.
c. A single-occupant toilet room with one water closet and one lavatory serving not more than two adjacent patient sleeping units shall be permitted provided that each patient sleeping unit has direct access to the toilet room and provision for privacy for the toilet room user is provided.
d. The occupant load for seasonal outdoor seating and entertainment areas shall be included when determining the minimum number of facilities required.
e. For business and mercantile classifications with an occupant load of 15 or fewer, service sinks shall not be required.
f. The required number and type of plumbing fixtures for outdoor public swimming pools shall be in accordance with Section 609 of the *International Swimming Pool and Spa Code*.

403.1.1 Fixture calculations. To determine the occupant load of each sex, the total occupant load shall be divided in half. To determine the required number of fixtures, the fixture ratio or ratios for each fixture type shall be applied to the occupant load of each sex in accordance with Table 403.1. Fractional numbers resulting from applying the fixture ratios of Table 403.1 shall be rounded up to the next whole number. For calculations involving multiple *occupancies*, such fractional numbers for each *occupancy* shall first be summed and then rounded up to the next whole number.

> **Exception:** The total occupant load shall not be required to be divided in half where *approved* statistical data indicates a distribution of the sexes of other than 50 percent of each sex.

403.1.2 Single-user toilet facility and bathing room fixtures. The plumbing fixtures located in single-user toilet facilities and bathing rooms, including family or assisted-use toilet and bathing rooms that are required by Section 1109.2.1 of the *International Building Code*, shall contribute toward the total number of required plumbing fixtures for a building or tenant space. Single-user toilet facilities and bathing rooms, and family or assisted-use toilet rooms and bathing rooms shall be identified for use by either sex.

403.1.3 Lavatory distribution. Where two or more toilet rooms are provided for each sex, the required number of lavatories shall be distributed proportionately to the required number of water closets.

403.2 Separate facilities. Where plumbing fixtures are required, separate facilities shall be provided for each sex.

Exceptions:

1. Separate facilities shall not be required for dwelling units and sleeping units.
2. Separate facilities shall not be required in structures or tenant spaces with a total occupant load, including both employees and customers, of 15 or fewer.
3. Separate facilities shall not be required in mercantile *occupancies* in which the maximum occupant load is 100 or fewer.
4. Separate facilities shall not be required in business occupancies in which the maximum occupant load is 25 or fewer.

403.2.1 Family or assisted-use toilet facilities serving as separate facilities. Where a building or tenant space requires a separate toilet facility for each sex and each toilet facility is required to have only one water closet, two family or assisted-use toilet facilities shall be permitted to serve as the required separate facilities. Family or assisted-use toilet facilities shall not be required to be identified for exclusive use by either sex as required by Section 403.4.

403.3 Employee and public toilet facilities. For structures and tenant spaces intended for *public* utilization, customers, patrons and visitors shall be provided with *public* toilet facilities. Employees associated with structures and tenant spaces shall be provided with toilet facilities. The number of plumbing fixtures located within the required toilet facilities shall be provided in accordance with Section 403 for all users. Employee toilet facilities shall be either separate or combined employee and *public* toilet facilities.

> **Exception:** *Public* toilet facilities shall not be required for:
> 1. Parking garages operated without parking attendants.
> 2. Structures and tenant spaces intended for quick transactions, including takeout, pickup and drop-off, having a public access area less than or equal to 300 square feet (28 m^2).

403.3.1 Access. The route to the *public* toilet facilities required by Section 403.3 shall not pass through kitchens, storage rooms or closets. Access to the required facilities shall be from within the building or from the exterior of the building. Routes shall comply with the accessibility requirements of the *International Building Code*. The public shall have access to the required toilet facilities at all times that the building is occupied.

403.3.2 Prohibited toilet room location. Toilet rooms shall not open directly into a room used for the preparation of food for service to the public.

403.3.3 Location of toilet facilities in occupancies other than malls. In occupancies other than covered and open mall buildings, the required *public* and employee toilet facilities shall be located not more than one story above or below the space required to be provided with toilet facilities, and the path of travel to such facilities shall not exceed a distance of 500 feet (152 m).

> **Exception:** The location and maximum distances of travel to required employee facilities in factory and industrial *occupancies* are permitted to exceed that required by this section, provided that the location and maximum distance of travel are *approved*.

403.3.4 Location of toilet facilities in malls. In covered and open mall buildings, the required *public* and employee toilet facilities shall be located not more than one story above or below the space required to be provided with toilet facilities, and the path of travel to such facilities shall not exceed a distance of 300 feet (91 m). In mall buildings, the required facilities shall be based on total square footage within a covered mall building or within the perimeter line of an open mall building, and facilities shall be installed in each individual store or in a central toilet area located in accordance with this section. The maximum distance of travel to central toilet facilities in mall buildings shall be measured from the main entrance of any store or tenant space. In mall buildings, where employees' toilet facilities are not provided in the individual store, the maximum distance of travel shall be measured from the employees' work area of the store or tenant space.

403.3.5 Pay facilities. Where pay facilities are installed, such facilities shall be in excess of the required minimum facilities. Required facilities shall be free of charge.

403.3.6 Door locking. Where a toilet room is provided for the use of multiple occupants, the egress door for the room shall not be lockable from the inside of the room. This section does not apply to family or assisted-use toilet rooms.

403.4 Signage. Required *public* facilities shall be provided with signs that designate the sex, as required by Section 403.2. Signs shall be readily visible and located near the entrance to each toilet facility. Signs for accessible toilet facilities shall comply with Section 1111 of the *International Building Code*.

403.4.1 Directional signage. Directional signage indicating the route to the required *public* toilet facilities shall be posted in a lobby, corridor, aisle or similar space, such that the sign can be readily seen from the main entrance to the building or tenant space.

403.5 Drinking fountain location. Drinking fountains shall not be required to be located in individual tenant spaces provided that *public* drinking fountains are located within a distance of travel of 500 feet (152 m) of the most remote location in the tenant space and not more than one story above or below the tenant space. Where the tenant space is in a covered or open mall, such distance shall not exceed 300 feet (91 m). Drinking fountains shall be located on an accessible route.

SECTION 404
ACCESSIBLE PLUMBING FACILITIES

404.1 Where required. Accessible plumbing facilities and fixtures shall be provided in accordance with the *International Building Code*.

404.2 Accessible fixture requirements. Accessible plumbing fixtures shall be installed with the clearances, heights, spacings and arrangements in accordance with ICC A117.1.

404.3 Exposed pipes and surfaces. Water supply and drain pipes under accessible lavatories and sinks shall be covered or otherwise configured to protect against contact. Pipe coverings shall comply with ASME A112.18.9.

SECTION 405
INSTALLATION OF FIXTURES

405.1 Water supply protection. The supply lines and fittings for every plumbing fixture shall be installed so as to prevent backflow.

405.2 Access for cleaning. Plumbing fixtures shall be installed so as to afford easy access for cleaning both the fixture and the area around the fixture.

405.3 Setting. Fixtures shall be set level and in proper alignment with reference to adjacent walls.

405.3.1 Water closets, urinals, lavatories and bidets. A water closet, urinal, lavatory or bidet shall not be set closer than 15 inches (381 mm) from its center to any side wall, partition, vanity or other obstruction. Where partitions or other obstructions do not separate adjacent fixtures, fixtures shall not be set closer than 30 inches (762 mm) center to center between adjacent fixtures. There shall be not less than a 21-inch (533 mm) clearance in front of a water closet, urinal, lavatory or bidet to any wall, fixture or door. Water closet compartments shall be not less than 30 inches (762 mm) in width and not less than 60 inches (1524 mm) in depth for floor-mounted water closets and not less than 30 inches (762 mm) in width and 56 inches (1422 mm) in depth for wall-hung water closets.

> **Exception:** An accessible children's water closet shall be set not closer than 12 inches (305 mm) from its center to the required partition or to the wall on one side.

405.3.2 Public lavatories. In employee and *public* toilet rooms, the required lavatory shall be located in the same room as the required water closet.

405.3.3 Location of fixtures and piping. Piping, fixtures or equipment shall not be located in such a manner as to interfere with the normal operation of windows, doors or other means of egress openings.

405.3.4 Water closet compartment. Each water closet utilized by the *public* or employees shall occupy a separate compartment with walls or partitions and a door enclosing the fixtures to ensure privacy.

> **Exceptions:**
> 1. Water closet compartments shall not be required in a single-occupant toilet room with a lockable door.
> 2. Toilet rooms located in child day care facilities and containing two or more water closets shall be permitted to have one water closet without an enclosing compartment.
> 3. This provision is not applicable to toilet areas located within Group I-3 housing areas.

405.3.5 Urinal partitions. Each urinal utilized by the *public* or employees shall occupy a separate area with walls or partitions to provide privacy. The horizontal dimension between walls or partitions at each urinal shall be not less than 30 inches (762 mm). The walls or partitions shall begin at a height not greater than 12 inches (305 mm) from and extend not less than 60 inches (1524 mm) above the finished floor surface. The walls or partitions shall extend from the wall surface at each side of the urinal not less than 18 inches (457 mm) or to a point not less than 6 inches (152 mm) beyond the outermost front lip of the urinal measured from the finished backwall surface, whichever is greater.

> **Exceptions:**
> 1. Urinal partitions shall not be required in a single-occupant or family/assisted-use toilet room with a lockable door.
> 2. Toilet rooms located in child day care facilities and containing two or more urinals shall be permitted to have one urinal without partitions.

405.4 Floor and wall drainage connections. Connections between the drain and floor outlet plumbing fixtures shall be made with a floor flange or a waste connector and sealing gasket. The waste connector and sealing gasket joint shall comply with the joint tightness test of ASME A112.4.3 and shall be installed in accordance with the manufacturer's instructions. The flange shall be attached to the drain and anchored to the structure. Connections between the drain and wall-hung water closets shall be made with an *approved* extension nipple or horn adaptor. The water closet shall be bolted to the hanger with corrosion-resistant bolts or screws. Joints shall be sealed with an *approved* elastomeric gasket, flange-to-fixture connection complying with ASME A112.4.3 or an *approved* setting compound.

405.4.1 Floor flanges. Floor flanges for water closets or similar fixtures shall be not less than 0.125 inch (3.2 mm) thick for copper alloy, 0.25 inch (6.4 mm) thick for plastic and 0.25 inch (6.4 mm) thick and not less than a 2-inch (51 mm) caulking depth for cast iron or galvanized malleable iron.

Floor flanges of hard lead shall weigh not less than 1 pound, 9 ounces (0.7 kg) and shall be composed of lead alloy with not less than 7.75-percent antimony by weight. Closet screws and bolts shall be of copper alloy. Flanges shall be secured to the building structure with corrosion-resistant screws or bolts.

405.4.2 Securing floor outlet fixtures. Floor outlet fixtures shall be secured to the floor or floor flanges by screws or bolts of corrosion-resistant material.

405.4.3 Securing wall-hung water closet bowls. Wall-hung water closet bowls shall be supported by a concealed metal carrier that is attached to the building structural members so that strain is not transmitted to the closet connector or any other part of the plumbing system. The carrier shall conform to ASME A112.6.2.

405.5 Plumbing fixtures with a pumped waste. Plumbing fixtures with a pumped waste shall comply with ASME A112.3.4/CSA B45.9. The plumbing fixture with a pumped waste shall be installed in accordance with the manufacturer's instructions.

405.6 Water-tight joints. Joints formed where fixtures come in contact with walls or floors shall be sealed.

405.7 Plumbing in mental health centers. In mental health centers, pipes or traps shall not be exposed, and fixtures shall be bolted through walls.

405.8 Design of overflows. Where any fixture is provided with an overflow, the waste shall be designed and installed so that standing water in the fixture will not rise in the overflow when the stopper is closed, and water will not remain in the overflow when the fixture is empty.

405.8.1 Connection of overflows. The overflow from any fixture shall discharge into the drainage system on the inlet or fixture side of the trap.

Exception: The overflow from a flush tank serving a water closet or urinal shall discharge into the fixture served.

405.9 Slip joint connections. Slip joints shall be made with an *approved* elastomeric gasket and shall only be installed on the trap outlet, trap inlet and within the trap seal. Fixtures with concealed slip-joint connections shall be provided with an *access* panel or utility space not less than 12 inches (305 mm) in its smallest dimension or other *approved* arrangement so as to provide *access* to the slip joint connections for inspection and repair.

405.10 Design and installation of plumbing fixtures. Integral fixture fitting mounting surfaces on manufactured plumbing fixtures or plumbing fixtures constructed on site shall meet the design requirements of ASME A112.19.2/CSA B45.1 or ASME A112.19.3/CSA B45.4.

SECTION 406
AUTOMATIC CLOTHES WASHERS

406.1 Water connection. The water supply to an automatic clothes washer shall be protected against backflow by an *air gap* that is integral with the machine or a backflow preventer shall be installed in accordance with Section 608. *Air gaps* shall comply with ASME A112.1.2 or A112.1.3.

406.2 Waste connection. The waste from an automatic clothes washer shall discharge through an *air break* into a standpipe in accordance with Section 802.3.3 or into a laundry sink. The trap and *fixture drain* for an automatic clothes washer standpipe shall be not less than 2 inches (51 mm) in diameter. The *fixture drain* for the standpipe serving an automatic clothes washer shall connect to a 3-inch (76 mm) or larger diameter fixture *branch* or *stack*. Automatic clothes washers that discharge by gravity shall be permitted to drain to a waste receptor or an *approved* trench drain.

SECTION 407
BATHTUBS

407.1 Approval. Bathtubs shall conform to ASME A112.19.1/CSA B45.2, ASME A112.19.2/CSA B45.1, ASME A112.19.3/CSA B45.4 or CSA B45.5/IAPMO Z124.

407.2 Bathtub waste outlets and overflows. Bathtubs shall be equipped with a waste outlet and an overflow outlet. The outlets shall be connected to waste tubing or piping not less than $1^1/_2$ inches (38 mm) in diameter. The waste outlet shall be equipped with a water-tight stopper.

407.3 Glazing. Windows and doors within a bathtub enclosure shall conform to the safety glazing requirements of the *International Building Code*.

407.4 Bathtub enclosure. Doors in a bathtub enclosure shall conform to ASME A112.19.15.

SECTION 408
BIDETS

408.1 Approval. Bidets shall conform to ASME A112.19.2/CSA B45.1.

408.2 Water connection. The water supply to a bidet shall be protected against backflow by an *air gap* or backflow preventer in accordance with Section 608.14.1, 608.14.2, 608.14.3, 608.14.5 or 608.14.6.

408.3 Bidet water temperature. The discharge water temperature from a bidet fitting shall be limited to not greater than 110°F (43°C) by a water-temperature-limiting device conforming to ASSE 1070/ASME A112.1070/CSA B125.70 or CSA B125.3.

SECTION 409
DISHWASHING MACHINES

409.1 Approval. Commercial dishwashing machines shall conform to ASSE 1004 and NSF 3. Residential dishwashers shall conform to NSF 184.

409.2 Water connection. The water supply to a dishwashing machine shall be protected against backflow by an *air gap* that is integral with the machine or a backflow preventer shall be installed in accordance with Section 608. *Air gaps* shall comply with ASME A112.1.2 or A112.1.3.

409.3 Waste connection. The waste connection of a commercial dishwashing machine shall comply with Section 802.1.6.

409.4 Residential dishwasher waste connection. The waste connection of a residential dishwasher shall connect directly to a wye branch fitting on the tailpiece of the kitchen sink, directly to the dishwasher connection of a food waste disposer, or through an air break to a standpipe. The waste line of a residential dishwasher shall rise and be securely fastened to the underside of the sink rim or counter top.

SECTION 410
DRINKING FOUNTAINS

410.1 Approval. Drinking fountains shall conform to ASME A112.19.1/CSA B45.2 or ASME A112.19.2/CSA B45.1 and *water coolers* shall conform to ASHRAE 18. Drinking fountains, *water coolers* and *water dispensers* shall conform to NSF 61, Section 9. Electrically operated, refrigerated drinking *water coolers* and *water dispensers* shall be listed and labeled in accordance with UL 399.

410.2 Small occupancies. Drinking fountains shall not be required for an occupant load of 15 or fewer.

[BE] 410.3 High and low drinking fountains. Where drinking fountains are required, not fewer than two drinking fountains shall be provided. One drinking fountain shall comply with the requirements for people who use a wheelchair and one drinking fountain shall comply with the requirements for standing persons.

Exceptions:

1. A single drinking fountain with two separate spouts that complies with the requirements for people who use a wheelchair and standing persons shall be permitted to be substituted for two separate drinking fountains.
2. Where drinking fountains are primarily for children's use, the drinking fountains for people using wheelchairs shall be permitted to comply with the children's provisions in ICC A117.1 and drinking fountains for standing children shall be permitted to provide the spout at 30 inches (762 mm) minimum above the floor.

410.4 Substitution. Where restaurants provide drinking water in a container free of charge, drinking fountains shall not be required in those restaurants. In other occupancies where drinking fountains are required, *water dispensers* shall be permitted to be substituted for not more than 50 percent of the required number of drinking fountains.

410.5 Prohibited location. Drinking fountains, *water coolers* and *water dispensers* shall not be installed in public restrooms.

SECTION 411
EMERGENCY SHOWERS AND EYEWASH STATIONS

411.1 Approval. Emergency showers and eyewash stations shall conform to ISEA Z358.1.

411.2 Waste connection. Waste connections shall not be required for emergency showers and eyewash stations.

411.3 Water supply. Where hot and cold water is supplied to an emergency shower or eyewash station, the temperature of the water supply shall only be controlled by a temperature actuated mixing valve complying with ASSE 1071.

SECTION 412
FAUCETS AND FIXTURE FITTINGS

412.1 Approval. Faucets and fixture fittings shall conform to ASME A112.18.1/CSA B125.1. Faucets and fixture fittings that supply drinking water for human ingestion shall conform to the requirements of NSF 61, Section 9. Flexible water connectors exposed to continuous pressure shall conform to the requirements of Section 605.6.

412.1.1 Faucets and supply fittings. Faucets and supply fittings shall conform to the water consumption requirements of Section 604.4.

412.1.2 Waste fittings. Waste fittings shall conform to ASME A112.18.2/CSA B125.2, ASTM F409 or to one of the standards listed in Tables 702.1 and 702.4 for aboveground drainage and vent pipe and fittings.

412.2 Hand showers. Hand-held showers shall conform to ASME A112.18.1/CSA B125.1. Hand-held showers shall provide backflow protection in accordance with ASME A112.18.1/CSA B125.1 or shall be protected against backflow by a device complying with ASME A112.18.3.

412.3 Individual shower valves. Individual shower and tub-shower combination valves shall be balanced-pressure, thermostatic or combination balanced-pressure/thermostatic valves that conform to the requirements of ASSE 1016/ASME A112.1016/CSA B125.16 or ASME A112.18.1/CSA B125.1 and shall be installed at the point of use. Shower and tub-shower combination valves required by this section shall be equipped with a means to limit the maximum setting of the valve to 120°F (49°C), which shall be field adjusted in accordance with the manufacturer's instructions. In-line thermostatic valves shall not be utilized for compliance with this section.

412.4 Multiple (gang) showers. Multiple (gang) showers supplied with a single-tempered water supply pipe shall have the water supply for such showers controlled by an *approved* automatic temperature control mixing valve that conforms to ASSE 1069 or CSA B125.3, or each shower head shall be individually controlled by a balanced-pressure, thermostatic or combination balanced-pressure/thermostatic valve that conforms to ASSE 1016/ASME A112.1016/CSA B125.16 or ASME A112.18.1/CSA B125.1 and is installed at the point of use. Such valves shall be equipped with a means to limit the maximum setting of the valve to 120°F (49°C), which shall be field adjusted in accordance with the manufacturers' instructions.

412.5 Bathtub and whirlpool bathtub valves. The *hot water* supplied to bathtubs and whirlpool bathtubs shall be limited to not greater than 120°F (49°C) by a water-temperature limiting device that conforms to ASSE 1070/ASME A112.1070/CSA B125.70 or CSA B125.3, except where such protection is otherwise provided by a combination tub/shower valve in accordance with Section 412.3.

412.6 Hose-connected outlets. Faucets and fixture fittings with hose-connected outlets shall conform to ASME A112.18.3 or ASME A112.18.1/CSA B125.1.

412.7 Temperature-actuated, flow-reduction devices for individual fixture fittings. Temperature-actuated, flow-reduction devices, where installed for individual fixture fittings, shall conform to ASSE 1062. A temperature-actuated, flow-reduction device shall be an *approved* method for limiting the water temperature to not greater than 120° F (49° C) at the outlet of a faucet or fixture fitting. Such devices shall not be used alone as a substitute for the balanced-pressure, thermostatic or combination shower valves required in Section 412.3 or as a substitute for bathtub or whirlpool tub water-temperature-limiting valves required in Section 412.5.

412.8 Transfer valves. Deck-mounted bath/shower transfer valves containing an integral atmospheric vacuum breaker shall conform to the requirements of ASME A112.18.1/CSA B125.1.

412.9 Water closet personal hygiene devices. Personal hygiene devices integral to water closets or water closet seats shall conform to the requirements of ASME A112.4.2/CSA B45.16.

412.10 Head shampoo sink faucets. Head shampoo sink faucets shall be supplied with hot water that is limited to not more than 120°F (49°C) by a water-temperature-limiting device that conforms to ASSE 1070/ASME A112.1070/CSA B125.70. Each faucet shall have integral check valves to prevent crossover flow between the hot and cold water supply connections.

SECTION 413
FLOOR AND TRENCH DRAINS

413.1 Approval. Floor drains shall conform to ASME A112.3.1, ASME A112.6.3 or CSA B79. Trench drains shall comply with ASME A112.6.3.

413.2 Floor drains. Floor drains shall have removable strainers. The floor drain shall be constructed so that the drain is capable of being cleaned. *Access* shall be provided to the drain inlet. Ready *access* shall be provided to floor drains.

> **Exception:** Floor drains serving refrigerated display cases shall be provided with *access*.

413.3 Size of floor drains. Floor drains shall have a drain outlet not less than 2 inches (51 mm) in diameter.

413.4 Public laundries and central washing facilities. In public coin-operated laundries and in the central washing facilities of multiple-family dwellings, the rooms containing automatic clothes washers shall be provided with floor drains located to readily drain the entire floor area. Such drains shall have an outlet of not less than 3 inches (76 mm) in diameter.

SECTION 414
FLOOR SINKS

414.1 Approval. Sanitary floor sinks shall conform to the requirements of ASME A112.6.7.

SECTION 415
FLUSHING DEVICES FOR
WATER CLOSETS AND URINALS

415.1 Flushing devices required. Each water closet, urinal, clinical sink and any plumbing fixture that depends on trap siphonage to discharge the fixture contents to the drainage system shall be provided with a flushometer valve, flushometer tank or a flush tank designed and installed to supply water in quantity and rate of flow to flush the contents of the fixture, cleanse the fixture and refill the fixture trap.

> **415.1.1 Separate for each fixture.** A flushing device shall not serve more than one fixture.

415.2 Flushometer valves and tanks. Flushometer valves and tanks shall comply with ASSE 1037/ASME A112.1037/CSA B125.37 or CSA B125.3. Vacuum breakers on flushometer valves shall conform to the performance requirements of ASSE 1001 or CSA B64.1.1. *Access* shall be provided to vacuum breakers. Flushometer valves shall be of the water conservation type and shall not be used where the water pressure is lower than the minimum required for normal operation. When operated, the valve shall automatically complete the cycle of operation, opening fully and closing positively under the water supply pressure. Each flushometer valve shall be provided with a means for regulating the flow through the valve. The trap seal to the fixture shall be automatically refilled after each flushing cycle.

415.3 Flush tanks. Flush tanks equipped for manual flushing shall be controlled by a device designed to refill the tank after each discharge and to shut off completely the water flow to the tank when the tank is filled to operational capacity. The trap seal to the fixture shall be automatically refilled after each flushing. The water supply to flush tanks equipped for automatic flushing shall be controlled with a timing device or sensor control devices.

> **415.3.1 Fill valves.** Flush tanks shall be equipped with an antisiphon fill valve conforming to ASSE 1002/ASME A112.1002/CSA B125.12 or CSA B125.3. The fill valve backflow preventer shall be located not less than 1 inch (25 mm) above the full opening of the overflow pipe.
>
> **415.3.2 Overflows in flush tanks.** Flush tanks shall be provided with overflows discharging to the water closet or urinal connected thereto and shall be sized to prevent flooding the tank at the maximum rate at which the tanks are supplied with water according to the manufacturer's

FIXTURES, FAUCETS AND FIXTURE FITTINGS

design conditions. The opening of the overflow pipe shall be located above the *flood level rim* of the water closet or urinal or above a secondary overflow in the flush tank.

415.3.3 Sheet copper. Sheet copper utilized for flush tank linings shall conform to ASTM B152 and shall not weigh less than 10 ounces per square foot (0.03 kg/m^2).

415.3.4 Access required. All parts in a flush tank shall be provided with *access* for repair and replacement.

415.4 Flush pipes and fittings. Flush pipes and fittings shall be of nonferrous material and shall conform to ASME A112.19.5/CSA B45.15.

SECTION 416
FOOD WASTE DISPOSER UNITS

416.1 Approval. Domestic food waste disposers shall conform to ASSE 1008 and shall be listed and labeled in accordance with UL 430. Food waste disposers shall not increase the *drainage fixture unit* load on the sanitary drainage system.

416.2 Domestic food waste disposer waste outlets. Domestic food waste disposers shall be connected to a drain of not less than $1^1/_2$ inches (38 mm) in diameter.

416.3 Commercial food waste disposer waste outlets. Commercial food waste disposers shall be connected to a drain not less than $1^1/_2$ inches (38 mm) in diameter. Commercial food waste disposers shall be connected and trapped separately from any other fixtures or sink compartments.

416.4 Water supply required. Food waste disposers shall be provided with a supply of cold water. The water supply shall be protected against backflow by an *air gap* or backflow preventer in accordance with Section 608.

SECTION 417
GARBAGE CAN WASHERS

417.1 Water connection. The water supply to a garbage can washer shall be protected against backflow by an *air gap* or a backflow preventer in accordance with Section 608.13.1, 608.13.2, 608.13.3, 608.13.5, 608.13.6 or 608.13.8.

417.2 Waste connection. Garbage can washers shall be trapped separately. The receptacle receiving the waste from the washer shall have a removable basket or strainer to prevent the discharge of large particles into the drainage system.

SECTION 418
LAUNDRY TRAYS

418.1 Approval. Laundry trays shall conform to ASME A112.19.1/CSA B45.2, ASME A112.19.2/CSA B45.1, ASME A112.19.3/CSA B45.4 or CSA B45.5/IAPMO Z124.

418.2 Waste outlet. Each compartment of a laundry tray shall be provided with a waste outlet not less than $1^1/_2$ inches (38 mm) in diameter and a strainer or crossbar to restrict the clear opening of the waste outlet.

SECTION 419
LAVATORIES

419.1 Approval. Lavatories shall conform to ASME A112.19.1/CSA B45.2, ASME A112.19.2/CSA B45.1, ASME A112.19.3/CSA B45.4 or CSA B45.5/IAPMO Z124. Group wash-up equipment shall conform to the requirements of Section 402. Every 20 inches (508 mm) of rim space shall be considered as one lavatory.

419.2 Cultured marble lavatories. Cultured marble vanity tops with an integral lavatory shall conform to CSA B45.5/IAPMO Z124.

419.3 Lavatory waste outlets. Lavatories shall have waste outlets not less than $1^1/_4$ inches (32 mm) in diameter. A strainer, pop-up stopper, crossbar or other device shall be provided to restrict the clear opening of the waste outlet.

419.4 Moveable lavatory systems. Moveable lavatory systems shall comply with ASME A112.19.12.

419.5 Tempered water for public hand-washing facilities. *Tempered water* shall be delivered from lavatories and group wash fixtures located in public toilet facilities provided for customers, patrons and visitors. *Tempered water* shall be delivered through an *approved* water-temperature limiting device that conforms to ASSE 1070/ASME A112.1070/CSA B125.70 or CSA B125.3.

SECTION 420
MANUAL FOOD AND BEVERAGE DISPENSING EQUIPMENT

420.1 Approval. Manual food and beverage dispensing equipment shall conform to the requirements of NSF 18.

SECTION 421
SHOWERS

421.1 Approval. Prefabricated showers and shower compartments shall conform to ASME A112.19.2/CSA B45.1 or CSA B45.5/IAPMO Z124. Shower valves for individual showers shall conform to the requirements of Section 412.3.

421.2 Water supply riser. Water supply risers from the shower valve to the shower head outlet, whether exposed or concealed, shall be attached to the structure. The attachment to the structure shall be made by the use of support devices designed for use with the specific piping material or by fittings anchored with screws.

421.3 Shower waste outlet. Waste outlets serving showers shall be not less than $1^1/_2$ inches (38 mm) in diameter and, for other than waste outlets in bathtubs, shall have removable strainers not less than 3 inches (76 mm) in diameter with strainer openings not less than $^1/_4$ inch (6.4 mm) in least dimension. Where each shower space is not provided with an individual waste outlet, the waste outlet shall be located and the floor pitched so that waste from one shower does not flow over the floor area serving another shower. Waste outlets shall be fastened to the waste pipe in an *approved* manner.

421.4 Shower compartments. Shower compartments shall be not less than 900 square inches (0.58 m^2) in interior cross-sectional area. Shower compartments shall be not less than 30 inches (762 mm) in least dimension as measured from the finished interior dimension of the compartment, exclusive of fixture valves, showerheads, soap dishes and safety grab bars or rails. Except as required in Section 404, the minimum required area and dimension shall be measured from the finished interior dimension at a height equal to the top of the threshold and at a point tangent to its centerline and shall be continued to a height not less than 70 inches (1778 mm) above the shower drain outlet.

Exception: Shower compartments having not less than 25 inches (635 mm) in minimum dimension measured from the finished interior dimension of the compartment, provided that the shower compartment has not less than 1,300 square inches (0.838 m^2) of cross-sectional area.

421.4.1 Floor and wall area. Bathtub floors, shower floors, wall areas above built-in tubs that have installed shower heads and walls in shower compartments shall be constructed of smooth, corrosion-resistant and nonabsorbent waterproof materials. Wall materials shall extend to a height of not less than 6 feet (1829 mm) above the room floor level, and not less than 70 inches (1778 mm) above the drain of the tub or shower. Such walls shall form a water-tight joint with each other and with either the tub or shower floor.

421.4.2 Access. The shower compartment access and egress opening shall have a clear and unobstructed finished width of not less than 22 inches (559 mm). Shower compartments required to be designed in conformance to accessibility provisions shall comply with Section 404.1.

421.5 Shower floors or receptors. Floor surfaces shall be constructed of impervious, noncorrosive, nonabsorbent and waterproof materials.

421.5.1 Support. Floors or receptors under shower compartments shall be laid on, and supported by, a smooth and structurally sound base.

421.5.2 Shower lining. Floors under shower compartments, except where prefabricated receptors have been provided, shall be lined and made watertight utilizing material complying with Sections 421.5.2.1 through 421.5.2.6. Such liners shall turn up on all sides not less than 2 inches (51 mm) above the finished threshold level. Liners shall be recessed and fastened to an *approved* backing so as not to occupy the space required for wall covering, and shall not be nailed or perforated at any point less than 1 inch (25 mm) above the finished threshold. Liners shall be pitched one-fourth unit vertical in 12 units horizontal (2-percent slope) and shall be sloped toward the fixture drains and be securely fastened to the waste outlet at the seepage entrance, making a water-tight joint between the liner and the outlet. The completed liner shall be tested in accordance with Section 312.9.

Exceptions:
1. Floor surfaces under shower heads provided for rinsing laid directly on the ground are not required to comply with this section.
2. Where a sheet-applied, load-bearing, bonded, waterproof membrane is installed as the shower lining, the membrane shall not be required to be recessed.

421.5.2.1 PVC sheets. Plasticized polyvinyl chloride (PVC) sheets shall meet the requirements of ASTM D4551. Sheets shall be joined by solvent welding in accordance with the manufacturer's installation instructions.

421.5.2.2 Chlorinated polyethylene (CPE) sheets. Nonplasticized chlorinated polyethylene sheet shall meet the requirements of ASTM D4068. The liner shall be joined in accordance with the manufacturer's installation instructions.

421.5.2.3 Sheet lead. Sheet lead shall weigh not less than 4 pounds per square foot (19.5 kg/m^2) and shall be coated with an asphalt paint or other *approved* coating. The lead sheet shall be insulated from conducting substances other than the connecting drain by 15-pound (6.80 kg) asphalt felt or an equivalent. Sheet lead shall be joined by burning.

421.5.2.4 Sheet copper. Sheet copper shall conform to ASTM B152 and shall weigh not less than 12 ounces per square foot (3.7 kg/m^2). The copper sheet shall be insulated from conducting substances other than the connecting drain by 15-pound (6.80 kg) asphalt felt or an equivalent. Sheet copper shall be joined by brazing or soldering.

421.5.2.5 Sheet-applied, load-bearing, bonded, waterproof membranes. Sheet-applied, load-bearing, bonded, waterproof membranes shall meet requirements of TCNA A118.10 and shall be applied in accordance with the manufacturer's installation instructions.

421.5.2.6 Liquid-type, trowel-applied, load-bearing, bonded waterproof materials. Liquid-type, trowel-applied, load-bearing, bonded waterproof materials shall meet the requirements of TCNA A118.10 and shall be applied in accordance with the manufacturer's instructions.

421.6 Glazing. Windows and doors within a shower enclosure shall conform to the safety glazing requirements of the *International Building Code*.

SECTION 422
SINKS

422.1 Approval. Sinks shall conform to ASME A112.19.1/CSA B45.2, ASME A112.19.2/CSA B45.1, ASME A112.19.3/CSA B45.4 or CSA B45.5/IAPMO Z124.

422.2 Sink waste outlets. Sinks shall be provided with waste outlets having a diameter not less than 1$^1/_2$ inches (38 mm). A strainer or crossbar shall be provided to restrict the clear opening of the waste outlet.

422.3 Moveable sink systems. Moveable sink systems shall comply with ASME A112.19.12.

SECTION 423
SPECIALTY PLUMBING FIXTURES

423.1 Water connections. Baptisteries, ornamental and lily pools, aquariums, ornamental fountain basins, swimming pools, and similar constructions, where provided with water supplies, shall be protected against backflow in accordance with Section 608.

423.2 Approval. Specialties requiring water and waste connections shall be submitted for approval.

423.3 Footbaths and pedicure baths. The water supplied to specialty plumbing fixtures, such as pedicure chairs having an integral foot bathtub and footbaths, shall be limited to not greater than 120°F (49°C) by a water-temperature-limiting device that conforms to ASSE 1070/ASME A112.1070/CSA B125.70 or CSA B125.3.

SECTION 424
URINALS

424.1 Approval. Urinals shall conform to ASME A112.19.2/CSA B45.1, ASME A112.19.19 or CSA B45.5/IAPMO Z124. Urinals shall conform to the water consumption requirements of Section 604.4. Water-supplied urinals shall conform to the hydraulic performance requirements of ASME A112.19.2/CSA B45.1 or CSA B45.5/IAPMO Z124.

424.2 Substitution for water closets. In each bathroom or toilet room, urinals shall not be substituted for more than 67 percent of the required water closets in assembly and educational *occupancies*. Urinals shall not be substituted for more than 50 percent of the required water closets in all other *occupancies*.

SECTION 425
WATER CLOSETS

425.1 Approval. Water closets shall conform to the water consumption requirements of Section 604.4 and shall conform to ASME A112.19.2/CSA B45.1, ASME A112.19.3/CSA B45.4 or CSA B45.5/IAPMO Z124. Water closets shall conform to the hydraulic performance requirements of ASME A112.19.2/CSA B45.1. Water closet tanks shall conform to ASME A112.19.2/CSA B45.1, ASME A112.19.3/CSA B45.4 or CSA B45.5/IAPMO Z124. Electro-hydraulic water closets shall comply with ASME A112.19.2/CSA B45.1. Water closets equipped with a dual flushing device shall comply with ASME A112.19.14.

425.2 Water closets for public or employee toilet facilities. Water closet bowls for *public* or employee toilet facilities shall be of the elongated type.

425.3 Water closet seats. Water closets shall be equipped with seats of smooth, nonabsorbent material. Seats of water closets provided for *public* or employee toilet facilities shall be of the hinged open-front type. Integral water closet seats shall be of the same material as the fixture. Water closet seats shall be sized for the water closet bowl type.

425.4 Water closet connections. A 4-inch by 3-inch (102 mm by 76 mm) closet bend shall be acceptable. Where a 3-inch (76 mm) bend is utilized on water closets, a 4-inch by 3-inch (102 mm by 76 mm) flange shall be installed to receive the fixture horn.

SECTION 426
WHIRLPOOL BATHTUBS

426.1 Approval. Whirlpool bathtubs shall comply with ASME A112.19.7/CSA B45.10 and shall be listed and labeled in accordance with UL 1795.

426.2 Installation. Whirlpool bathtubs shall be installed and tested in accordance with the manufacturer's instructions. The pump shall be located above the weir of the fixture trap.

426.3 Drain. The pump drain and circulation piping shall be sloped to drain the water in the volute and the circulation piping when the whirlpool bathtub is empty.

426.4 Suction fittings. Suction fittings for whirlpool bathtubs shall comply with ASME A112.19.7/CSA B45.10.

426.5 Access to pump. *Access* shall be provided to circulation pumps in accordance with the fixture or pump manufacturer's installation instructions. Where the manufacturer's instructions do not specify the location and minimum size of field-fabricated access openings, an opening not less than 12 inches by 12 inches (305 mm by 305 mm) shall be installed to provide *access* to the circulation pump. Where pumps are located more than 2 feet (609 mm) from the access opening, an opening not less than 18 inches by 18 inches (457 mm by 457 mm) shall be installed. A door or panel shall be permitted to close the opening. In all cases, the access opening shall be unobstructed and of the size necessary to permit the removal and replacement of the circulation pump.

426.6 Whirlpool enclosure. Doors within a whirlpool enclosure shall conform to ASME A112.19.15.

CHAPTER 5
WATER HEATERS

User note:

About this chapter: *Chapter 5 contains regulations concerning the safety of water heating units and hot water storage tanks. Heated (hot or tempered) potable water is needed for plumbing fixtures that are associated with handwashing, bathing, culinary activities and building maintenance. Heated water is commonly stored in large pressurized storage tanks that must be protected against explosion by pressure and temperature relief valves specified in this chapter. This chapter also covers the access requirements to water heaters and hot water storage tanks to allow for the maintenance and replacement of that equipment.*

SECTION 501
GENERAL

501.1 Scope. The provisions of this chapter shall govern the materials, design and installation of water heaters and the related safety devices and appurtenances.

501.2 Water heater as space heater. Where a combination potable water heating and space heating system requires water for space heating at temperatures greater than 140°F (60°C), a master thermostatic mixing valve complying with ASSE 1017 shall be provided to limit the water supplied to the potable *hot water* distribution system to a temperature of 140°F (60°C) or less. The potability of the water shall be maintained throughout the system.

501.3 Drain valves. Drain valves for emptying shall be installed at the bottom of each tank-type water heater and hot water storage tank. The drain valve inlet shall be not less than $^3/_4$-inch (19 mm) nominal iron pipe size and the outlet shall be provided with male garden hose threads.

501.4 Location. Water heaters and storage tanks shall be located and connected so as to provide *access* for observation, maintenance, servicing and replacement.

501.5 Water heater labeling. Water heaters shall be third-party certified.

501.6 Water temperature control in piping from tankless heaters. The temperature of water from tankless water heaters shall be not greater than 140°F (60°C) where intended for domestic uses. This provision shall not supersede the requirement for protective shower valves in accordance with Section 412.3.

501.7 Pressure marking of storage tanks. Storage tanks and water heaters installed for domestic hot water shall have the maximum allowable working pressure clearly and indelibly stamped in the metal or marked on a plate welded thereto or otherwise permanently attached. Such markings shall be in a position with *access* on the outside of the tank so as to make inspection or reinspection readily possible.

501.8 Temperature controls. Hot water supply systems shall be equipped with automatic temperature controls capable of adjustments from the lowest to the highest acceptable temperature settings for the intended temperature operating range.

SECTION 502
INSTALLATION

502.1 General. Water heaters shall be installed in accordance with the manufacturer's instructions. Oil-fired water heaters shall conform to the requirements of this code and the *International Mechanical Code*. Electric water heaters shall conform to the requirements of this code and provisions of NFPA 70. Gas-fired water heaters shall conform to the requirements of the *International Fuel Gas Code*. Solar thermal water heating systems shall conform to the requirements of the *International Mechanical Code* and ICC 900/SRCC 300.

>**502.1.1 Elevation and protection.** Elevation of water heater ignition sources and mechanical damage protection requirements for water heaters shall be in accordance with the *International Mechanical Code* and the *International Fuel Gas Code*.

502.2 Rooms used as a plenum. Water heaters using solid, liquid or gas fuel shall not be installed in a room containing air-handling machinery where such room is used as a plenum.

502.3 Water heaters installed in attics. Attics containing a water heater shall be provided with an opening and unobstructed passageway large enough to allow removal of the water heater. The passageway shall be not less than 30 inches (762 mm) in height and 22 inches (559 mm) in width and not more than 20 feet (6096 mm) in length when measured along the centerline of the passageway from the opening to the water heater. The passageway shall have continuous solid flooring not less than 24 inches (610 mm) in width. A level service space not less than 30 inches (762 mm) in length and 30 inches (762 mm) in width shall be present at the front or service side of the water heater. The clear access opening dimensions shall be not less than 20 inches by 30 inches (508 mm by 762 mm) where such dimensions are large enough to allow removal of the water heater.

502.4 Seismic supports. Where earthquake loads are applicable in accordance with the *International Building Code*, water heater supports shall be designed and installed for the seismic forces in accordance with the *International Building Code*.

502.5 Clearances for maintenance and replacement. Appliances shall be provided with *access* for inspection, service, repair and replacement without disabling the function of a fire-resistance-rated assembly or removing permanent construction, other appliances or any other piping or ducts not connected to the appliance being inspected, serviced, repaired or replaced. A level working space not less than 30 inches in length and 30 inches in width (762 mm by 762 mm) shall be provided in front of the control side to service an appliance.

SECTION 503
CONNECTIONS

503.1 Cold water line valve. The cold water *branch* line from the main water supply line to each hot water storage tank or water heater shall be provided with a valve, located near the equipment and serving only the hot water storage tank or water heater. The valve shall not interfere or cause a disruption of the cold water supply to the remainder of the cold water system. The valve shall be provided with *access* on the same floor level as the water heater served.

503.2 Water circulation. The method of connecting a circulating water heater to the tank shall provide proper circulation of water through the water heater. The pipe or tubes required for the installation of appliances that will draw from the water heater or storage tank shall comply with the provisions of this code for material and installation.

SECTION 504
SAFETY DEVICES

504.1 Antisiphon devices. An *approved* means, such as a cold water "dip" tube with a hole at the top or a vacuum relief valve installed in the cold water supply line above the top of the heater or tank, shall be provided to prevent siphoning of any storage water heater or tank.

504.2 Vacuum relief valve. Bottom fed water heaters and bottom fed tanks connected to water heaters shall have a vacuum relief valve installed. The vacuum relief valve shall comply with ANSI Z21.22.

504.3 Shutdown. A means for disconnecting an electric hot water supply system from its energy supply shall be provided in accordance with NFPA 70. A separate valve shall be provided to shut off the energy fuel supply to all other types of hot water supply systems.

504.4 Relief valve. Storage water heaters operating above atmospheric pressure shall be provided with an *approved*, self-closing (levered) pressure relief valve and temperature relief valve or combination thereof. The relief valve shall conform to ANSI Z21.22. The relief valve shall not be used as a means of controlling thermal expansion.

504.4.1 Installation. Such valves shall be installed in the shell of the water heater tank. Temperature relief valves shall be so located in the tank as to be actuated by the water in the top 6 inches (152 mm) of the tank served. For installations with separate storage tanks, the *approved*, self-closing (levered) pressure relief valve and temperature relief valve or combination thereof conforming to ANSI Z21.22 valves shall be installed on both the storage water heater and storage tank. There shall not be a check valve or shutoff valve between a relief valve and the heater or tank served.

504.5 Relief valve approval. Temperature and pressure relief valves, or combinations thereof, and energy cutoff devices shall bear the label of an *approved* agency and shall have a temperature setting of not more than 210°F (99°C) and a pressure setting not exceeding the tank or water heater manufacturer's rated working pressure or 150 psi (1035 kPa), whichever is less. The relieving capacity of each pressure relief valve and each temperature relief valve shall equal or exceed the heat input to the water heater or storage tank.

504.6 Requirements for discharge piping. The discharge piping serving a pressure relief valve, temperature relief valve or combination thereof shall:

1. Not be directly connected to the drainage system.
2. Discharge through an *air gap* located in the same room as the water heater.
3. Not be smaller than the diameter of the outlet of the valve served and shall discharge full size to the *air gap*.
4. Serve a single relief device and shall not connect to piping serving any other relief device or equipment.
5. Discharge to the floor, to the pan serving the water heater or storage tank, to a waste receptor or to the outdoors.
6. Discharge in a manner that does not cause personal injury or structural damage.
7. Discharge to a termination point that is readily observable by the building occupants.
8. Not be trapped.
9. Be installed so as to flow by gravity.
10. Terminate not more than 6 inches (152 mm) above and not less than two times the discharge pipe diameter above the floor or *flood level rim* of the waste receptor.
11. Not have a threaded connection at the end of such piping.
12. Not have valves or tee fittings.
13. Be constructed of those materials listed in Section 605.4 or materials tested, rated and *approved* for such use in accordance with ASME A112.4.1.
14. Be one nominal size larger than the size of the relief valve outlet, where the relief valve discharge piping is installed with insert fittings. The outlet end of such tubing shall be fastened in place.

504.7 Required pan. Where a storage tank-type water heater or a hot water storage tank is installed in a location where water leakage from the tank will cause damage, the tank shall be installed in a pan constructed of one of the following:

1. Galvanized steel or aluminum of not less than 0.0236 inch (0.6010 mm) in thickness.
2. Plastic not less than 0.036 inch (0.9 mm) in thickness.
3. Other *approved* materials.

A plastic pan shall not be installed beneath a gas-fired water heater.

504.7.1 Pan size and drain. The pan shall be not less than $1^1/_2$ inches (38 mm) in depth and shall be of sufficient size and shape to receive all dripping or condensate from the tank or water heater. The pan shall be drained by an indirect waste pipe having a diameter of not less than $^3/_4$ inch (19 mm). Piping for safety pan drains shall be of those materials listed in Table 605.4.

504.7.2 Pan drain termination. The pan drain shall extend full size and terminate over a suitably located indirect waste receptor or floor drain or extend to the exterior of the building and terminate not less than 6 inches (152 mm) and not more than 24 inches (610 mm) above the adjacent ground surface. Where a pan drain was not previously installed, a pan drain shall not be required for a replacement water heater installation.

SECTION 505
INSULATION

[E] **505.1 Unfired vessel insulation.** Unfired hot water storage tanks shall be insulated to R-12.5 (h • ft^2 • °F)/Btu (R-2.2 m^2 • K/W).

CHAPTER 6

WATER SUPPLY AND DISTRIBUTION

User note:

About this chapter: Many plumbing fixtures require a supply of potable water. Other fixtures could be supplied with nonpotable water such as reclaimed water. Chapter 6 covers the requirements for water distribution piping systems to and within buildings. The regulations include the types of materials and the connection methods for such systems. The prevention of backflow of contaminated or polluted water into any potable water system is critical for protection of users of potable water. This chapter regulates the assemblies, devices and methods that are used for this purpose.

SECTION 601
GENERAL

601.1 Scope. This chapter shall govern the materials, design and installation of water supply systems, both hot and cold, for utilization in connection with human occupancy and habitation and shall govern the installation of individual water supply systems.

601.2 Solar energy utilization. Solar energy systems used for heating potable water or using an independent medium for heating potable water shall comply with the applicable requirements of this code. The use of solar energy shall not compromise the requirements for cross connection or protection of the potable water supply system required by this code.

601.3 Existing piping used for grounding. Existing metallic water service piping used for electrical grounding shall not be replaced with nonmetallic pipe or tubing until other *approved* means of grounding is provided.

601.4 Tests. The potable water distribution system shall be tested in accordance with Section 312.5.

601.5 Rehabilitation of piping systems. Where pressure piping systems are rehabilitated using an epoxy lining system, such lining system shall comply with ASTM F2831.

SECTION 602
WATER REQUIRED

602.1 General. Structures equipped with plumbing fixtures and utilized for human occupancy or habitation shall be provided with a potable supply of water in the amounts and at the pressures specified in this chapter.

602.2 Potable water required. Only potable water shall be supplied to plumbing fixtures that provide water for drinking, bathing or culinary purposes, or for the processing of food, medical or pharmaceutical products. Unless otherwise provided in this code, potable water shall be supplied to all plumbing fixtures.

602.3 Individual water supply. Where a potable public water supply is not available, individual sources of potable water supply shall be utilized.

602.3.1 Sources. Dependent on geological and soil conditions and the amount of rainfall, individual water supplies are of the following types: drilled well, driven well, dug well, bored well, spring, stream or cistern. Surface bodies of water and land cisterns shall not be sources of individual water supply unless properly treated by *approved* means to prevent contamination. Individual water supplies shall be constructed and installed in accordance with the applicable state and local laws. Where such laws do not address all of the requirements set forth in NGWA-01, individual water supplies shall comply with NGWA-01 for those requirements not addressed by state and local laws.

602.3.2 Minimum quantity. The combined capacity of the source and storage in an individual water supply system shall supply the fixtures with water at rates and pressures as required by this chapter.

602.3.3 Water quality. Water from an individual water supply shall be *approved* as potable by the authority having jurisdiction prior to connection to the plumbing system.

602.3.4 Disinfection of system. After construction, the individual water supply system shall be purged of deleterious matter and disinfected in accordance with Section 610.

602.3.5 Pumps. Pumps shall be rated for the transport of potable water. Pumps in an individual water supply system shall be constructed and installed so as to prevent contamination from entering a potable water supply through the pump units. Pumps shall be sealed to the well casing or covered with a water-tight seal. Pumps shall be designed to maintain a prime and installed such that ready *access* is provided to the pump parts of the entire assembly for repairs.

602.3.5.1 Pump enclosure. The pump room or enclosure around a well pump shall be drained and protected from freezing by heating or other *approved* means. Where pumps are installed in basements, such pumps shall be mounted on a block or shelf not less than 18 inches (457 mm) above the basement floor. Well pits shall be prohibited.

SECTION 603
WATER SERVICE

603.1 Size of water service pipe. The water service pipe shall be sized to supply water to the structure in the quantities and at the pressures required in this code. The water service pipe shall be not less than $^3/_4$ inch (19.1 mm) in diameter.

603.2 Separation of water service and building sewer. Where water service piping is located in the same trench with

WATER SUPPLY AND DISTRIBUTION

the *building sewer*, such *sewer* shall be constructed of materials listed in Table 702.2. Where the *building sewer* piping is not constructed of materials listed in Table 702.2, the water service pipe and the *building sewer* shall be horizontally separated by not less than 5 feet (1524 mm) of undisturbed or compacted earth. The required separation distance shall not apply where a water service pipe crosses a *sewer* pipe, provided that the water service is sleeved to a point not less than 5 feet (1524 mm) horizontally from the *sewer* pipe centerline on both sides of such crossing. The sleeve shall be of pipe materials listed in Table 605.3, 702.2 or 702.3. The required separation distance shall not apply where the bottom of the water service pipe, located within 5 feet (1524 mm) of the *sewer,* is not less than 12 inches (305 mm) above the highest point of the top of the *building sewer.*

603.2.1 Water service near sources of pollution. Potable water service pipes shall not be located in, under or above cesspools, septic tanks, septic tank drainage fields or seepage pits. Where soil or ground water causes contaminated conditions for piping, analysis shall be required in accordance with Section 605.1.

SECTION 604
DESIGN OF BUILDING
WATER DISTRIBUTION SYSTEM

604.1 General. The design of the water distribution system shall conform to *accepted engineering practice*. Methods utilized to determine pipe sizes shall be *approved*.

604.2 System interconnection. At the points of interconnection between the hot and cold water supply piping systems and the individual fixtures, appliances or devices, provisions shall be made to prevent flow between such piping systems.

604.3 Water distribution system design criteria. The water distribution system shall be designed, and pipe sizes shall be selected such that under conditions of peak demand, the capacities at the fixture supply pipe outlets shall be not less than shown in Table 604.3. The minimum flow rate and flow pressure provided to fixtures and appliances not listed in Table 604.3 shall be in accordance with the manufacturer's installation instructions.

604.4 Maximum flow and water consumption. The maximum water consumption flow rates and quantities for all plumbing fixtures and fixture fittings shall be in accordance with Table 604.4.

Exceptions:

1. Blowout design water closets having a water consumption not greater than $3^1/_2$ gallons (13 L) per flushing cycle.
2. Vegetable sprays.
3. Clinical sinks having a water consumption not greater than $4^1/_2$ gallons (17 L) per flushing cycle.
4. Service sinks.
5. Emergency showers.

TABLE 604.3
WATER DISTRIBUTION SYSTEM DESIGN CRITERIA REQUIRED CAPACITY AT FIXTURE SUPPLY PIPE OUTLETS

FIXTURE SUPPLY OUTLET SERVING	FLOW RATE[a] (gpm)	FLOW PRESSURE (psi)
Bathtub, balanced-pressure, thermostatic or combination balanced-pressure/thermostatic mixing valve	4	20
Bidet, thermostatic mixing valve	2	20
Combination fixture	4	8
Dishwasher, residential	2.75	8
Drinking fountain	0.75	8
Laundry tray	4	8
Lavatory, private	0.8	8
Lavatory, private, mixing valve	0.8	8
Lavatory, public	0.4	8
Shower	2.5	8
Shower, balanced-pressure, thermostatic or combination balanced-pressure/thermostatic mixing valve	2.5[b]	20
Sillcock, hose bibb	5	8
Sink, residential	1.75	8
Sink, service	3	8
Urinal, valve	12	25
Water closet, blow out, flushometer valve	25	45
Water closet, flushometer tank	1.6	20
Water closet, siphonic, flushometer valve	25	35
Water closet, tank, close coupled	3	20
Water closet, tank, one piece	6	20

For SI: 1 pound per square inch = 6.895 kPa,
1 gallon per minute = 3.785 L/m.

a. For additional requirements for flow rates and quantities, see Section 604.4.
b. Where the shower mixing valve manufacturer indicates a lower flow rating for the mixing valve, the lower value shall be applied.

TABLE 604.4
MAXIMUM FLOW RATES AND CONSUMPTION FOR PLUMBING FIXTURES AND FIXTURE FITTINGS

PLUMBING FIXTURE OR FIXTURE FITTING	MAXIMUM FLOW RATE OR QUANTITY[b]
Lavatory, private	2.2 gpm at 60 psi
Lavatory, public (metering)	0.25 gallon per metering cycle
Lavatory, public (other than metering)	0.5 gpm at 60 psi
Shower head[a]	2.5 gpm at 80 psi
Sink faucet	2.2 gpm at 60 psi
Urinal	1.0 gallon per flushing cycle
Water closet	1.6 gallons per flushing cycle

For SI: 1 gallon = 3.785 L, 1 gallon per minute = 3.785 L/m,
1 pound per square inch = 6.895 kPa.

a. A hand-held shower spray is a shower head.
b. Consumption tolerances shall be determined from referenced standards.

604.5 Size of fixture supply. The minimum size of a fixture supply pipe shall be as shown in Table 604.5. The fixture supply pipe shall terminate not more than 30 inches (762 mm) from the point of connection to the fixture. A reduced-size flexible water connector installed between the supply pipe and the fixture shall be of an *approved* type. The supply pipe shall extend to the floor or wall adjacent to the fixture. The minimum size of individual distribution lines utilized in gridded or parallel water distribution systems shall be as shown in Table 604.5.

TABLE 604.5
MINIMUM SIZES OF FIXTURE WATER SUPPLY PIPES

FIXTURE	MINIMUM PIPE SIZE (inch)
Bathtubs[a] (60″ × 32″ and smaller)	1/2
Bathtubs[a] (larger than 60″ × 32″)	1/2
Bidet	3/8
Combination sink and tray	1/2
Dishwasher, domestic[a]	1/2
Drinking fountain	3/8
Hose bibbs	1/2
Kitchen sink[a]	1/2
Laundry, 1, 2 or 3 compartments[a]	1/2
Lavatory	3/8
Shower, single head[a]	1/2
Sinks, flushing rim	3/4
Sinks, service	1/2
Urinal, flush tank	1/2
Urinal, flushometer valve	3/4
Wall hydrant	1/2
Water closet, flush tank	3/8
Water closet, flushometer tank	3/8
Water closet, flushometer valve	1
Water closet, one piece[a]	1/2

For SI: 1 inch = 25.4 mm, 1 foot = 304.8 mm, 1 pound per square inch = 6.895 kPa.

a. Where the developed length of the distribution line is 50 feet or less, and the available pressure at the meter is 35 psi or greater, the minimum size of an individual distribution line supplied from a manifold and installed as part of a parallel water distribution system shall be one nominal tube size smaller than the sizes indicated.

604.6 Variable street pressures. Where street water main pressures fluctuate, the building water distribution system shall be designed for the minimum pressure available.

604.7 Inadequate water pressure. Wherever water pressure from the street main or other source of supply is insufficient to provide flow pressures at fixture outlets as required under Table 604.3, a water pressure booster system conforming to Section 606.5 shall be installed on the building water supply system.

604.8 Water pressure-reducing valve or regulator. Where water pressure within a building exceeds 80 psi (552 kPa) static, an *approved* water pressure-reducing valve conforming to ASSE 1003 or CSA B356 with strainer shall be installed to reduce the pressure in the building water distribution piping to not greater than 80 psi (552 kPa) static.

Exception: Service lines to sill cocks and outside hydrants, and main supply risers where pressure from the mains is reduced to 80 psi (552 kPa) or less at individual fixtures.

604.8.1 Valve design. The pressure-reducing valve shall be designed to remain open to permit uninterrupted water flow in case of valve failure.

604.8.2 Repair and removal. Water pressure-reducing valves, regulators and strainers shall be so constructed and installed as to permit repair or removal of parts without breaking a pipeline or removing the valve and strainer from the pipeline.

604.9 Water hammer. The flow velocity of the water distribution system shall be controlled to reduce the possibility of water hammer. A water-hammer arrestor shall be installed where *quick-closing valves* are utilized. Water-hammer arrestors shall be installed in accordance with the manufacturer's instructions. Water-hammer arrestors shall conform to ASSE 1010.

604.10 Gridded and parallel water distribution system manifolds. Hot water and cold water manifolds installed with gridded or parallel connected individual distribution lines to each fixture or fixture fitting shall be designed in accordance with Sections 604.10.1 through 604.10.3.

604.10.1 Manifold sizing. Hot water and cold water manifolds shall be sized in accordance with Table 604.10.1. The total gallons per minute is the demand of all outlets supplied.

TABLE 604.10.1
MANIFOLD SIZING

NOMINAL SIZE INTERNAL DIAMETER (inches)	MAXIMUM DEMAND (gpm)	
	Velocity at 4 feet per second	Velocity at 8 feet per second
1/2	2	5
3/4	6	11
1	10	20
1 1/4	15	31
1 1/2	22	44

For SI: 1 inch = 25.4 mm, 1 gallon per minute = 3.785 L/m, 1 foot per second = 0.305 m/s.

604.10.2 Valves. Individual fixture shutoff valves installed at the manifold shall be identified as to the fixture being supplied.

604.10.3 Access. *Access* shall be provided to manifolds with integral factory- or field-installed valves.

604.11 Individual pressure balancing in-line valves for individual fixture fittings. Where individual pressure balancing in-line valves for individual fixture fittings are installed, such valves shall comply with ASSE 1066. Such valves shall be installed in a location with access. The valves shall not be utilized alone as a substitute for the balanced pressure, thermostatic or combination shower valves required in Section 412.3.

SECTION 605
MATERIALS, JOINTS AND CONNECTIONS

605.1 Soil and ground water. The installation of a water service or water distribution pipe shall be prohibited in soil and ground water contaminated with solvents, fuels, organic compounds or other detrimental materials causing permeation, corrosion, degradation or structural failure of the piping material. Where detrimental conditions are suspected, a chemical analysis of the soil and ground water conditions shall be required to ascertain the acceptability of the water service or water distribution piping material for the specific installation. Where detrimental conditions exist, *approved* alternative materials or routing shall be required.

605.2 Lead content of water supply pipe and fittings. Pipe and pipe fittings, including valves and faucets, utilized in the water supply system shall have not more than 8-percent lead content.

605.2.1 Lead content of drinking water pipe and fittings. Pipe, pipe fittings, joints, valves, faucets and fixture fittings utilized to supply water for drinking or cooking purposes shall comply with NSF 372 and shall have a weighted average lead content of 0.25 percent or less.

605.3 Water service pipe. Water service pipe shall conform to NSF 61 and shall conform to one of the standards listed in Table 605.3. Water service pipe or tubing, installed underground and outside of the structure, shall have a working pressure rating of not less than 160 psi (1100 kPa) at 73.4°F (23°C). Where the water pressure exceeds 160 psi (1100 kPa), piping material shall have a working pressure rating not less than the highest available pressure. Water service piping materials not third-party certified for water distribution shall terminate at or before the full open valve located at the entrance to the structure. Ductile iron water service piping shall be cement mortar lined in accordance with AWWA C104/A21.4.

605.3.1 Dual check-valve-type backflow preventer. Dual check-valve backflow preventers installed on the water supply system shall comply with ASSE 1024 or CSA B64.6.

605.4 Water distribution pipe. Water distribution pipe and tubing shall conform to NSF 61 and shall conform to one of the standards listed in Table 605.4. Hot water distribution pipe and tubing shall have a pressure rating of not less than 100 psi (690 kPa) at 180°F (82°C).

605.5 Fittings. Pipe fittings shall be approved for installation with the piping material installed and shall comply with the applicable standards listed in Table 605.5. Pipe fittings utilized in water supply systems shall also comply with NSF 61. Ductile and gray iron pipe and pipe fittings utilized in water service piping systems shall be cement mortar lined in accordance with AWWA C104/A21.4.

605.5.1 Mechanically formed tee fittings. Mechanically extracted outlets shall have a height not less than three times the thickness of the branch tube wall.

TABLE 605.3
WATER SERVICE PIPE

MATERIAL	STANDARD
Acrylonitrile butadiene styrene (ABS) plastic pipe	ASTM D1527; ASTM D2282
Chlorinated polyvinyl chloride (CPVC) plastic pipe	ASTM D2846; ASTM F441; ASTM F442; CSA B137.6
Chlorinated polyvinyl chloride/aluminum/chlorinated polyvinyl chloride (CPVC/AL/CPVC)	ASTM F2855
Copper or copper-alloy pipe	ASTM B42; ASTM B302
Copper or copper-alloy tubing (Type K, WK, L, WL, M or WM)	ASTM B75; ASTM B88; ASTM B251; ASTM B447
Cross-linked polyethylene (PEX) plastic pipe and tubing	ASTM F876; AWWA C904; CSA B137.5
Cross-linked polyethylene/aluminum/cross-linked polyethylene (PEX-AL-PEX) pipe	ASTM F1281; ASTM F2262; CSA B137.10
Cross-linked polyethylene/aluminum/high-density polyethylene (PEX-AL-HDPE)	ASTM F1986
Ductile iron water pipe	AWWA C151/A21.51; AWWA C115/A21.15
Galvanized steel pipe	ASTM A53
Polyethylene (PE) plastic pipe	ASTM D2239; ASTM D3035; AWWA C901; CSA B137.1
Polyethylene (PE) plastic tubing	ASTM D2737; AWWA C901; CSA B137.1
Polyethylene/aluminum/polyethylene (PE-AL-PE) pipe	ASTM F1282; CSA B137.9
Polyethylene of raised temperature (PE-RT) plastic tubing	ASTM F2769; CSA B137.18
Polypropylene (PP) plastic pipe or tubing	ASTM F2389; CSA B137.11
Polyvinyl chloride (PVC) plastic pipe	ASTM D1785; ASTM D2241; ASTM D2672; CSA B137.3
Stainless steel pipe (Type 304/304L)	ASTM A312; ASTM A778
Stainless steel pipe (Type 316/316L)	ASTM A312; ASTM A778

TABLE 605.4
WATER DISTRIBUTION PIPE

MATERIAL	STANDARD
Chlorinated polyvinyl chloride (CPVC) plastic pipe and tubing	ASTM D2846; ASTM F441; ASTM F442; CSA B137.6
Chlorinated polyvinyl chloride/aluminum/chlorinated polyvinyl chloride (CPVC/AL/CPVC)	ASTM F2855
Copper or copper-alloy pipe	ASTM B42; ASTM B302; ASTM B43
Copper or copper-alloy tubing (Type K, WK, L, WL, M or WM)	ASTM B75; ASTM B88; ASTM B251; ASTM B447
Cross-linked polyethylene (PEX) plastic tubing	ASTM F876; CSA B137.5
Cross-linked polyethylene/aluminum/cross-linked polyethylene (PEX-AL-PEX) pipe	ASTM F1281; ASTM F2262; CSA B137.10
Cross-linked polyethylene/aluminum/high-density polyethylene (PEX-AL-HDPE)	ASTM F1986
Ductile iron pipe	AWWA C151/A21.51; AWWA C115/A21.15
Galvanized steel pipe	ASTM A53
Polyethylene/aluminum/polyethylene (PE-AL-PE) composite pipe	ASTM F1282
Polyethylene of raised temperature (PE-RT) plastic tubing	ASTM F2769; CSA B137.18
Polypropylene (PP) plastic pipe or tubing	ASTM F2389; CSA B137.11
Stainless steel pipe (Type 304/304L)	ASTM A312; ASTM A778
Stainless steel pipe (Type 316/316L)	ASTM A312; ASTM A778

605.5.1.1 Full flow assurance. Branch tubes shall not restrict the flow in the run tube. A dimple serving as a depth stop shall be formed in the branch tube to ensure that penetration into the collar is of the correct depth. For inspection purposes, a second dimple shall be placed $^1/_4$ inch (6.4 mm) above the first dimple. Dimples shall be aligned with the tube run.

605.5.1.2 Brazed joints. Mechanically formed tee fittings shall be brazed in accordance with Section 605.14.1.

605.6 Flexible water connectors. Flexible water connectors exposed to continuous pressure shall conform to ASME A112.18.6/CSA B125.6. *Access* shall be provided to all flexible water connectors.

605.7 Valves. Valves shall be compatible with the type of piping material installed in the system. Valves shall conform to one of the standards listed in Table 605.7 or shall be *approved*. Valves intended to supply drinking water shall meet the requirements of NSF 61.

605.8 Manufactured pipe nipples. Manufactured pipe nipples shall conform to one of the standards listed in Table 605.8.

TABLE 605.8
MANUFACTURED PIPE NIPPLES

MATERIAL	STANDARD
Copper, copper alloy, and chromium-plated	ASTM B687
Steel	ASTM A733

605.9 Prohibited joints and connections. The following types of joints and connections shall be prohibited:

1. Cement or concrete joints.
2. Joints made with fittings not *approved* for the specific installation.
3. Solvent-cement joints between different types of plastic pipe.
4. Saddle-type fittings.

605.10 ABS plastic. Joints between ABS plastic pipe and fittings shall comply with Sections 605.10.1 through 605.10.3.

605.10.1 Mechanical joints. Mechanical joints on water pipes shall be made with an elastomeric seal conforming to ASTM D3139. Mechanical joints shall only be installed in underground systems, unless otherwise *approved*. Joints shall be installed only in accordance with the manufacturer's instructions.

605.10.2 Solvent cementing. Joint surfaces shall be clean and free from moisture. Solvent cement that conforms to ASTM D2235 shall be applied to all joint surfaces. The joint shall be made while the cement is wet. Joints shall be made in accordance with ASTM D2235. Solvent-cement joints shall be permitted above or below ground.

605.10.3 Threaded joints. Threads shall conform to ASME B1.20.1. Where pipe is to be threaded, the pipe shall have a wall thickness of not less than Schedule 80. Pipe threads shall be made with dies specifically designed for plastic pipe. *Approved* thread lubricant or tape shall be applied on the male threads only.

605.11 Gray iron and ductile iron joints. Joints for gray and ductile iron pipe and fittings shall comply with AWWA C111/A21.11 and shall be installed in accordance with the manufacturer's instructions.

605.12 Copper pipe. Joints between copper or copper-alloy pipe and fittings shall comply with Sections 605.12.1 through 605.12.5.

WATER SUPPLY AND DISTRIBUTION

**TABLE 605.5
PIPE FITTINGS**

MATERIAL	STANDARD
Acrylonitrile butadiene styrene (ABS) plastic	ASTM D2468
Cast iron	ASME B16.4
Chlorinated polyvinyl chloride (CPVC) plastic	ASSE 1061; ASTM D2846; ASTM F437; ASTM F438; ASTM F439; CSA B137.6
Copper or copper alloy	ASME B16.15; ASME B16.18; ASME B16.22; ASME B16.26; ASME B16.51; ASSE 1061; ASTM F1476; ASTM F1548
Cross-linked polyethylene/aluminum/high-density polyethylene (PEX-AL-HDPE)	ASTM F1986
Fittings for cross-linked polyethylene (PEX) plastic tubing	ASSE 1061, ASTM F877; ASTM F1807; ASTM F1960; ASTM F2080; ASTM F2098, ASTM F2159; ASTM F2434; ASTM F2735; CSA B137.5
Fittings for polyethylene of raised temperature (PE-RT) plastic tubing	ASSE 1061, ASTM D3261; ASTM F1807; ASTM F2098; ASTM F2159; ASTM F2735; ASTM F2769; CSA B137.18
Gray iron and ductile iron	ASTM F1476; ASTM F1548; AWWA C110/A21.10; AWWA C153/A21.53
Insert fittings for polyethylene/aluminum/polyethylene (PE-AL-PE) and cross-linked polyethylene/aluminum/cross-linked polyethylene (PEX-AL-PEX)	ASTM F1974; ASTM F1281; ASTM F1282; CSA B137.9; CSA B137.10
Malleable iron	ASME B16.3
Metal (brass) insert fittings for polyethylene/aluminum/polyethylene (PE-AL-PE) and cross-linked polyethylene/aluminum/cross-linked polyethylene (PEX-AL-PEX)	ASTM F1974
Polyethylene (PE) plastic pipe	ASTM D2609; ASTM D2683; ASTM D3261; ASTM F1055; CSA B137.1
Polypropylene (PP) plastic pipe or tubing	ASTM F2389; CSA B137.11
Polyvinyl chloride (PVC) plastic	ASTM D2464; ASTM D2466; ASTM D2467; CSA B137.2; CSA B137.3
Stainless steel (Type 304/304L)	ASTM A312; ASTM A778; ASTM F1476; ASTM F1548
Stainless steel (Type 316/316L)	ASTM A312; ASTM A778; ASTM F1476; ASTM F1548
Steel	ASME B16.9; ASME B16.11; ASME B16.28; ASTM F1476; ASTM F1548

**TABLE 605.7
VALVES**

MATERIAL	STANDARD
Chlorinated polyvinyl chloride (CPVC) plastic	ASME A112.4.14; ASME A112.18.1/CSA B125.1; ASTM F1970; CSA B125.3; IAPMO Z1157; MSS SP-122
Copper or copper alloy	ASME A112.4.14; ASME A112.18.1/CSA B125.1; ASME B16.34; CSA B125.3; MSS SP-67; MSS SP-80; MSS SP-110; IAPMO Z1157; MSS SP-139
Cross-linked polyethylene (PEX) plastic	ASME A112.4.14; ASME A112.18.1/CSA B125.1; CSA B125.3; NSF 359; IAPMO Z1157
Gray iron and ductile iron	AWWA C500; AWWA C504; AWWA C507; MSS SP-67; MSS SP-70; MSS SP-71; MSS SP-72; MSS SP-78; IAPMO Z1157
Polypropylene (PP) plastic	ASME A112.4.14; ASTM F2389; IAPMO Z1157
Polyvinyl chloride (PVC) plastic	ASME A112.4.14; ASTM F1970; IAPMO Z1157; MSS SP-122

605.12.1 Brazed joints. Joint surfaces shall be cleaned. An *approved* flux shall be applied where required. The joint shall be brazed with a filler metal conforming to AWS A5.8.

605.12.2 Mechanical joints. Mechanical joints shall be installed in accordance with the manufacturer's instructions.

605.12.3 Solder joints. Solder joints shall be made in accordance with ASTM B828. Cut tube ends shall be reamed to the full inside diameter of the tube end. Joint surfaces shall be cleaned. A flux conforming to ASTM B813 shall be applied. The joint shall be soldered with a solder conforming to ASTM B32. The joining of water supply piping shall be made with *lead-free solder and fluxes*. "Lead free" shall mean a chemical composition equal to or less than 0.2-percent lead.

605.12.4 Threaded joints. Threads shall conform to ASME B1.20.1. Pipe-joint compound or tape shall be applied on the male threads only.

605.12.5 Welded joints. Joint surfaces shall be cleaned. The joint shall be welded with an *approved* filler metal.

605.13 Copper tubing. Joints between copper or copper-alloy tubing and fittings shall comply with Sections 605.13.1 through 605.13.7.

605.13.1 Brazed joints. Joint surfaces shall be cleaned. An *approved* flux shall be applied where required. The joint shall be brazed with a filler metal conforming to AWS A5.8.

605.13.2 Flared joints. Flared joints for water pipe shall be made by a tool designed for that operation.

605.13.3 Grooved and shouldered mechanical joints. Grooved and shouldered mechanical joints shall comply with ASTM F1476, shall be made with an *approved* elastomeric seal and shall be installed in accordance with the manufacturer's instructions. Such joints shall be exposed or concealed.

605.13.4 Mechanical joints. Mechanical joints shall be installed in accordance with the manufacturer's instructions.

605.13.5 Press-connect joints. Press-connect joints shall conform to one of the standards indicated in Table 605.5, and shall be installed in accordance with the manufacturer's instructions. Cut tube ends shall be reamed to the full inside diameter of the tube end. Joint surfaces shall be cleaned. The tube shall be fully inserted into the press-connect fitting. Press-connect joints shall be pressed with a tool certified by the manufacturer.

605.13.6 Solder joints. Solder joints shall be made in accordance with the methods of ASTM B828. Cut tube ends shall be reamed to the full inside diameter of the tube end. Joint surfaces shall be cleaned. A flux conforming to ASTM B813 shall be applied. The joint shall be soldered with a solder conforming to ASTM B32. The joining of water supply piping shall be made with *lead-free solder and flux*. "Lead free" shall mean a chemical composition equal to or less than 0.2-percent lead.

605.13.7 Push-fit joints. Push-fit joints shall conform to ASSE 1061 and shall be installed in accordance with the manufacturer's instructions.

605.14 CPVC plastic. Joints between CPVC plastic pipe and fittings shall comply with Sections 605.14.1 through 605.14.4.

605.14.1 Mechanical joints. Mechanical joints shall be installed in accordance with the manufacturer's instructions.

605.14.2 Solvent cementing. Joint surfaces shall be clean and free from moisture. Joints shall be made in accordance with the pipe manufacturer's installation instructions. Where such instructions require that a primer be used, the primer shall be applied to the joint surfaces and a solvent cement orange in color and conforming to ASTM F493 shall be applied to the joint surfaces. Where such instructions allow for a one-step solvent cement, yellow in color and conforming to ASTM F493, to be used, the joint surfaces shall not require application of a primer before the solvent cement is applied. The joint shall be made while the cement is wet and in accordance with ASTM D2846 or ASTM F493. Solvent cemented joints shall be permitted above or below ground.

605.14.3 Threaded joints. Threads shall conform to ASME B1.20.1. Where pipe is to be threaded, the pipe shall have a wall thickness of not less than Schedule 80. Pipe threads shall be made with dies specifically designed for plastic pipe. The pressure rating of threaded pipe shall be reduced by 50 percent. Thread-by-socket molded fittings shall be permitted. *Approved* thread lubricant or tape shall be applied on the male threads only.

605.14.4 Push-fit joints. Push-fit joints shall conform to ASSE 1061 and shall be installed in accordance with the manufacturer's instructions.

605.15 Chlorinated polyvinyl chloride/aluminum/chlorinated polyvinyl chloride (CPVC/AL/CPVC) pipe and tubing. Joints between CPVC/AL/CPVC plastic pipe or CPVC fittings shall comply with Sections 605.15.1 and 605.15.2.

605.15.1 Mechanical joints. Mechanical joints shall be installed in accordance with the manufacturer's instructions.

605.15.2 Solvent cementing. Joint surfaces shall be clean and free from moisture, and an approved primer shall be applied. Solvent cement, orange in color and conforming to ASTM F493, shall be applied to joint surfaces. The joint shall be made while the cement is wet, and in accordance with ASTM D2846 or ASTM F493. Solvent cement joints shall be permitted above or below ground.

> **Exception:** A primer is not required where all of the following conditions apply:
> 1. The solvent cement used is third-party certified as conforming to ASTM F493.
> 2. The solvent cement used is yellow in color.
> 3. The solvent cement is used only for joining $^1/_2$-inch (12.7 mm) through 2-inch-diameter (51

mm) CPVC/AL/CPVC pipe and CPVC fittings.

4. The CPVC fittings are manufactured in accordance with ASTM D2846.

605.16 PEX plastic. Joints between cross-linked polyethylene plastic tubing and fittings shall comply with Sections 605.16.1 through 605.16.3.

605.16.1 Flared joints. Flared pipe ends shall be made by a tool designed for that operation.

605.16.2 Mechanical joints. Mechanical joints shall be installed in accordance with the manufacturer's instructions. Fittings for cross-linked polyethylene (PEX) plastic tubing shall comply with the applicable standards listed in Table 605.5 and shall be installed in accordance with the manufacturer's instructions. PEX tubing shall be factory marked with the appropriate standards for the fittings that the PEX manufacturer specifies for use with the tubing.

605.16.3 Push-fit joints. Push-fit joints shall conform to ASSE 1061 and shall be installed in accordance with the manufacturer's instructions.

605.17 Steel. Joints between galvanized steel pipe and fittings shall comply with Sections 605.17.1 through 605.17.3.

605.17.1 Threaded joints. Threads shall conform to ASME B1.20.1. Pipe-joint compound or tape shall be applied on the male threads only.

605.17.2 Mechanical joints. Joints shall be made with an *approved* elastomeric seal. Mechanical joints shall be installed in accordance with the manufacturer's instructions.

605.17.3 Grooved and shouldered mechanical joints. Grooved and shouldered mechanical joints shall comply with ASTM F1476, shall be made with an *approved* elastomeric seal and shall be installed in accordance with the manufacturer's instructions. Such joints shall be exposed or concealed.

605.18 PE plastic. Joints between polyethylene plastic pipe or tubing and fittings shall comply with Sections 605.18.1 through 605.18.4.

605.18.1 Flared joints. Flared joints shall be permitted where so indicated by the pipe manufacturer. Flared joints shall be made by a tool designed for that operation.

605.18.2 Heat-fusion joints. Joint surfaces shall be clean and free from moisture. Joint surfaces shall be heated to melt temperature and joined. The joint shall be undisturbed until cool. Joints shall be made in accordance with ASTM D2657.

605.18.3 Mechanical joints. Mechanical joints shall be installed in accordance with the manufacturer's instructions.

605.18.4 Installation. Polyethylene pipe shall be cut square, with a cutter designed for plastic pipe. Except where joined by heat fusion, pipe ends shall be chamfered to remove sharp edges. Kinked pipe shall not be installed. The minimum pipe bending radius shall be not less than 30 pipe diameters, or the minimum coil radius, whichever is greater. Piping shall not be bent beyond straightening of the curvature of the coil. Bends shall be prohibited within 10 pipe diameters of any fitting or valve. Stiffener inserts installed with compression-type couplings and fittings shall not extend beyond the clamp or nut of the coupling or fitting.

605.19 Polypropylene (PP) plastic. Joints between PP plastic pipe and fittings shall comply with Section 605.19.1 or 605.19.2.

605.19.1 Heat-fusion joints. Heat-fusion joints for polypropylene pipe and tubing joints shall be installed with socket-type heat-fused polypropylene fittings, butt-fusion polypropylene fittings or electrofusion polypropylene fittings. Joint surfaces shall be clean and free from moisture. The joint shall be undisturbed until cool. Joints shall be made in accordance with ASTM F2389.

605.19.2 Mechanical and compression sleeve joints. Mechanical and compression sleeve joints shall be installed in accordance with the manufacturer's instructions.

605.20 Polyethylene/aluminum/polyethylene (PE-AL-PE) and cross-linked polyethylene/aluminum/cross-linked polyethylene (PEX-AL-PEX). Joints between PE-AL-PE or PEX-AL-PEX pipe and fittings shall comply with Section 605.20.1.

605.20.1 Mechanical joints. Mechanical joints shall be installed in accordance with the manufacturer's instructions. Fittings for PE-AL-PE and PEX-AL-PEX as described in ASTM F1974, ASTM F1281, ASTM F1282, CSA B137.9 and CSA B137.10 shall be installed in accordance with the manufacturer's instructions.

605.21 PVC plastic. Joints between PVC plastic pipe and fittings shall comply with Sections 605.21.1 through 605.21.3.

605.21.1 Mechanical joints. Mechanical joints on water pipe shall be made with an elastomeric seal conforming to ASTM D3139. Mechanical joints shall not be installed in above-ground systems unless otherwise *approved*. Joints shall be installed in accordance with the manufacturer's instructions.

605.21.2 Grooved and shouldered mechanical joints. Grooved and shouldered mechanical joints shall comply with ASTM F1476, shall be made with an *approved* elastomeric seal and shall be installed in accordance with the manufacturer's instructions. Such joints shall be exposed or concealed.

605.21.3 Solvent cementing. Joint surfaces shall be clean and free from moisture. A purple primer that conforms to ASTM F656 shall be applied. Solvent cement not purple in color and conforming to ASTM D2564 or CSA B137.3 shall be applied to all joint surfaces. The joint shall be made while the cement is wet and shall be in accordance with ASTM D2855. Solvent-cement joints shall be permitted above or below ground.

605.21.4 Threaded joints. Threads shall conform to ASME B1.20.1. Where pipe is to be threaded, the pipe shall have a wall thickness of not less than Schedule 80.

Pipe threads shall be made with dies specifically designed for plastic pipe. The pressure rating of threaded pipe shall be reduced by 50 percent. Thread-by-socket molded fittings shall be permitted. Approved thread lubricant or tape shall be applied on the male threads only.

605.22 Stainless steel. Joints between stainless steel pipe and fittings shall comply with Sections 605.22.1 through 605.22.3.

> **605.22.1 Mechanical joints.** Mechanical joints shall be installed in accordance with the manufacturer's instructions.
>
> **605.22.2 Welded joints.** Joint surfaces shall be cleaned. The joint shall be welded autogenously or with an *approved* filler metal as referenced in ASTM A312.
>
> **605.22.3 Grooved and shouldered mechanical joints.** Grooved and shouldered mechanical joints shall comply with ASTM F1476, shall be made with an *approved* elastomeric seal and shall be installed in accordance with the manufacturer's instructions. Such joints shall be exposed or concealed.

605.23 Joints between different materials. Joints between different piping materials shall be made with a mechanical joint of the compression or mechanical-sealing type, or shall be made in accordance with Section 605.23.1, 605.23.2 or 605.23.3. Connectors or adapters shall have an elastomeric seal conforming to ASTM F477. Joints shall be installed in accordance with the manufacturer's instructions.

> **605.23.1 Copper or copper-alloy tubing to galvanized steel pipe.** Joints between copper pipe or tubing and galvanized steel pipe shall be made with a copper-alloy or dielectric fitting or a dielectric union conforming to ASSE 1079. The copper tubing shall be soldered to the fitting in an *approved* manner, and the fitting shall be screwed to the threaded pipe.
>
> **605.23.2 Plastic pipe or tubing to other piping material.** Joints between different types of plastic pipe or between plastic pipe and other piping material shall be made with *approved* adapters or transition fittings.
>
> **605.23.3 Stainless steel.** Joints between stainless steel and different piping materials shall be made with a mechanical joint of the compression or mechanical sealing type or a dielectric fitting or a dielectric union conforming to ASSE 1079.

605.24 PE-RT plastic. Joints between polyethylene of raised temperature plastic tubing and fittings shall be in accordance with Section 605.24.1.

> **605.24.1 Mechanical joints.** Mechanical joints shall be installed in accordance with the manufacturer's instructions. Fittings for polyethylene of raised temperature plastic tubing shall comply with the applicable standards indicated in Table 605.5 and shall be installed in accordance with the manufacturer's instructions. Polyethylene of raised temperature plastic tubing shall be factory marked with the applicable standards for the fittings that the manufacturer of the tubing specifies for use with the tubing.

SECTION 606
INSTALLATION OF THE BUILDING WATER DISTRIBUTION SYSTEM

606.1 Location of full-open valves. *Full-open valves* shall be installed in the following locations:

1. On the building water service pipe from the public water supply near the curb.
2. On the water distribution supply pipe at the entrance into the structure.
3. On the discharge side of every water meter.
4. On the base of every water riser pipe in occupancies other than multiple-family residential *occupancies* that are two stories or less in height and in one- and two-family residential *occupancies*.
5. On the top of every water down-feed pipe in *occupancies* other than one- and two-family residential *occupancies*.
6. On the entrance to every water supply pipe to a dwelling unit, except where supplying a single fixture equipped with individual stops.
7. On the water supply pipe to a gravity or pressurized water tank.
8. On the water supply pipe to every water heater.

606.2 Location of shutoff valves. Shutoff valves shall be installed in the following locations:

1. On the fixture supply to each plumbing fixture other than bathtubs and showers in one- and two-family residential *occupancies*, and other than in individual sleeping units that are provided with unit shutoff valves in hotels, motels, boarding houses and similar *occupancies*.
2. On the water supply pipe to each sillcock.
3. On the water supply pipe to each appliance or mechanical equipment.

606.3 Access to valves. *Access* shall be provided to all *full-open valves* and shutoff valves.

606.4 Valve identification. Service and hose bibb valves shall be identified. Other valves installed in locations that are not adjacent to the fixture or appliance shall be identified, indicating the fixture or appliance served.

606.5 Water pressure booster systems. Water pressure booster systems shall be provided as required by Sections 606.5.1 through 606.5.10.

> **606.5.1 Water pressure booster systems required.** Where the water pressure in the public water main or individual water supply system is insufficient to supply the minimum pressures and quantities specified in this code, the supply shall be supplemented by an elevated water tank, a hydropneumatic pressure booster system or a water pressure booster pump installed in accordance with Section 606.5.5.
>
> **606.5.2 Support.** Water supply tanks shall be supported in accordance with the *International Building Code*.

WATER SUPPLY AND DISTRIBUTION

606.5.3 Covers. Water supply tanks shall be covered to keep out unauthorized persons, dirt and vermin. The covers of gravity tanks shall be vented with a return bend vent pipe with an area not less than the area of the down-feed riser pipe, and the vent shall be screened with a corrosion-resistant screen of not less than 16 by 20 mesh per inch (630 by 787 mesh per m).

606.5.4 Overflows for water supply tanks. A gravity or suction water supply tank shall be provided with an overflow with a diameter not less than that shown in Table 606.5.4. The overflow outlet shall discharge at a point not less than 6 inches (152 mm) above the roof or roof drain; floor or floor drain; or over an open water-supplied fixture. The overflow outlet shall be covered with a corrosion-resistant screen of not less than 16 by 20 mesh per inch (630 by 787 mesh per m) and by $^1/_4$-inch (6.4 mm) hardware cloth or shall terminate in a horizontal angle seat check valve. Drainage from overflow pipes shall be directed so as not to freeze on roof walks.

TABLE 606.5.4
SIZES FOR OVERFLOW PIPES FOR WATER SUPPLY TANKS

MAXIMUM CAPACITY OF WATER SUPPLY LINE TO TANK (gpm)	DIAMETER OF OVERFLOW PIPE (inches)
0 – 50	2
50 – 150	$2^1/_2$
150 – 200	3
200 – 400	4
400 – 700	5
700 – 1,000	6
Over 1,000	8

For SI: 1 inch = 25.4 mm, 1 gallon per minute = 3.785 L/m.

606.5.5 Low-pressure cutoff required on booster pumps. A low-pressure cutoff shall be installed on all booster pumps in a water pressure booster system to prevent creation of a vacuum or negative pressure on the suction side of the pump when a positive pressure of 10 psi (68.94 kPa) or less occurs on the suction side of the pump.

606.5.6 Potable water inlet control and location. Potable water inlets to gravity tanks shall be controlled by a fill valve or other automatic supply valve installed so as to prevent the tank from overflowing. The inlet shall be terminated so as to provide an *air gap* not less than 4 inches (102 mm) above the overflow.

606.5.7 Tank drain pipes. A drain pipe with a valve shall be provided at the lowest point of each tank to permit emptying of the tank. The tank drain pipe shall discharge as required for overflow pipes and shall not be smaller in size than specified in Table 606.5.7.

TABLE 606.5.7
SIZE OF DRAIN PIPES FOR WATER TANKS

TANK CAPACITY (gallons)	DRAIN PIPE (inches)
Up to 750	1
751 to 1,500	$1^1/_2$
1,501 to 3,000	2
3,001 to 5,000	$2^1/_2$
5,000 to 7,500	3
Over 7,500	4

For SI: 1 inch = 25.4 mm, 1 gallon = 3.785 L.

606.5.8 Prohibited location of potable supply tanks. Potable water gravity tanks or manholes of potable water pressure tanks shall not be located directly under any soil or waste piping or any source of contamination.

606.5.9 Pressure tanks, vacuum relief. Water pressure tanks shall be provided with a vacuum relief valve at the top of the tank that will operate up to a maximum water pressure of 200 psi (1380 kPa) and up to a maximum temperature of 200°F (93°C). The size of such vacuum relief valve shall be not less than $^1/_2$ inch (12.7 mm).

Exception: This section shall not apply to pressurized captive air diaphragm/bladder tanks.

606.5.10 Pressure relief for tanks. Every pressure tank in a hydropneumatic pressure booster system shall be protected with a pressure relief valve. The pressure relief valve shall be set at a maximum pressure equal to the rating of the tank. The relief valve shall be installed on the supply pipe to the tank or on the tank. The relief valve shall discharge by gravity to a safe place of disposal.

606.6 Water supply system test. Upon completion of a section of or the entire water supply system, the system, or portion completed, shall be tested in accordance with Section 312.

606.7 Labeling of water distribution pipes in bundles. Where water distribution piping is bundled at installation, each pipe in the bundle shall be identified using stenciling or commercially available pipe labels. The identification shall indicate the pipe contents and the direction of flow in the pipe. The interval of the identification markings on the pipe shall not exceed 25 feet (7620 mm). There shall be not less than one identification label on each pipe in each room, space or story.

SECTION 607
HOT WATER SUPPLY SYSTEM

607.1 Where required. In residential *occupancies*, *hot water* shall be supplied to plumbing fixtures and equipment utilized for bathing, washing, culinary purposes, cleansing, laundry or building maintenance. In nonresidential *occupancies*, *hot water* shall be supplied for culinary purposes, cleansing, laundry or building maintenance purposes. In nonresidential *occupancies*, *hot water* or *tempered water* shall be supplied for bathing and washing purposes.

607.1.1 Temperature limiting means. A thermostat control for a water heater shall not serve as the temperature limiting means for the purposes of complying with the requirements of this code for maximum allowable *hot* or *tempered water* delivery temperature at fixtures.

607.1.2 Tempered water temperature control. *Tempered water* shall be supplied through a water temperature limiting device that conforms to ASSE 1070/ASME A112.1070/CSA B125.70 and shall limit the *tempered water* to not greater than 110°F (43°C). This provision shall not supersede the requirement for protective shower valves in accordance with Section 412.3.

607.2 Hot or tempered water supply to fixtures. The *developed length* of *hot* or *tempered water* piping, from the source of hot water to the fixtures that require *hot* or *tempered water*, shall not exceed 50 feet (15 240 mm). Recirculating system piping and heat-traced piping shall be considered to be sources of *hot* or *tempered water*.

607.2.1 Circulation systems and heat trace systems for maintaining heated water temperature in distribution systems. For Group R2, R3 and R4 occupancies that are three stories or less in height above grade plane, the installation of heated water circulation and temperature maintenance systems shall be in accordance with Section R403.5.1 of the *International Energy Conservation Code*. For other than Group R2, R3 and R4 occupancies that are three stories or less in height above grade plane, the installation of heated water circulation and heat trace systems shall be in accordance with Section C404.6 of the *International Energy Conservation Code*.

607.2.1.1 Pump controls for hot water storage systems. The controls on pumps that circulate water between a water heater and a storage tank for heated water shall limit operation of the pump from heating cycle startup to not greater than 5 minutes after the end of the cycle.

607.2.1.2 Demand recirculation controls for distribution systems. A water distribution system having one or more recirculation pumps that pump water from a heated water supply pipe back to the heated water source through a cold water supply pipe shall be a demand recirculation water system. Pumps shall have controls that comply with both of the following:

1. The control shall start the pump upon receiving a signal from the action of a user of a fixture or appliance, sensing the presence of a user of a fixture, or sensing the flow of hot or tempered water to a fixture fitting or appliance.

2. The control shall limit the temperature of the water entering the cold water piping to 104°F (40°C).

607.2.2 Piping for recirculation systems having master thermostatic valves. Where a thermostatic mixing valve is used in a system with a hot water recirculating pump, the *hot water* or *tempered water* return line shall be routed to the cold water inlet pipe of the water heater and the cold water inlet pipe or the hot water return connection of the thermostatic mixing valve.

607.3 Thermal expansion control. Where a storage water heater is supplied with cold water that passes through a check valve, pressure reducing valve or backflow preventer, a thermal expansion control device shall be connected to the water heater cold water supply pipe at a point that is downstream of all check valves, pressure reducing valves and backflow preventers. Thermal expansion tanks shall be sized in accordance with the tank manufacturer's instructions and shall be sized such that the pressure in the water distribution system shall not exceed that required by Section 604.8.

607.4 Flow of hot water to fixtures. Fixture fittings, faucets and diverters shall be installed and adjusted so that the flow of hot water from the fittings corresponds to the left-hand side of the fixture fitting.

Exception: Shower and tub/shower mixing valves conforming to ASSE 1016/ASME A112.1016/CSA B125.16 or ASME A112.18.1/CSA B125.1, where the flow of hot water corresponds to the markings on the device.

[E] 607.5 Insulation of piping. For other than Group R2, R3 and R4 occupancies that are three stories or less in height above grade plane, piping to the inlet of a water heater and piping conveying water heated by a water heater shall be insulated in accordance with Section C404.4 of the *International Energy Conservation Code*. For Group R2, R3 and R4 occupancies that are three stories or less in height above grade plane, piping to the inlet of a water heater and piping conveying water heated by a water heater shall be insulated in accordance with Section R403.5.3 of the *International Energy Conservation Code*.

SECTION 608
PROTECTION OF POTABLE WATER SUPPLY

608.1 General. A potable water supply system shall be designed, installed and maintained in such a manner so as to prevent contamination from nonpotable liquids, solids or gases being introduced into the potable water supply through cross connections or any other piping connections to the system. Backflow preventer applications shall conform to Table 608.1, except as specifically stated in Sections 608.2 through 608.17.10.

608.2 Plumbing fixtures. The supply lines and fittings for plumbing fixtures shall be installed so as to prevent backflow. Plumbing fixture fittings shall provide backflow protection in accordance with ASME A112.18.1/CSA B125.1.

608.3 Devices, appurtenances, appliances and apparatus. Devices, appurtenances, appliances and apparatus intended to serve some special function, such as sterilization, distillation, processing, cooling, or storage of ice or foods, and that connect to the water supply system, shall be provided with protection against backflow and contamination of the water supply system.

608.3.1 Special equipment, water supply protection. The water supply for hospital fixtures shall be protected against backflow with a reduced pressure principle backflow prevention assembly, an atmospheric or spill-resistant vacuum breaker assembly, or an *air gap*. Vacuum breakers for bedpan washer hoses shall not be located less than 5 feet (1524 mm) above the floor. Vacuum breakers for hose connections in health care or laboratory areas shall be not less than 6 feet (1829 mm) above the floor.

608.4 Potable water handling and treatment equipment. Water pumps, filters, softeners, tanks and other appliances and devices that handle or treat potable water to be supplied to the potable water distribution system shall be located to prevent contamination from entering the appliances and devices. Overflow, relief valve and waste discharge pipes from such appliances and devices shall terminate through an air gap.

608.5 Water service piping. Water service piping shall be protected in accordance with Sections 603.2 and 603.2.1.

608.6 Chemicals and other substances. Chemicals and other substances that produce either toxic conditions, taste, odor or discoloration in a potable water system shall not be introduced into, or utilized in, such systems.

608.7 Cross connection control. Cross connections shall be prohibited, except where *approved* backflow prevention assemblies, backflow prevention devices or other means or methods are installed to protect the potable water supply.

608.7.1 Private water supplies. Cross connections between a private water supply and a potable public supply shall be prohibited.

608.8 Valves and outlets prohibited below grade. Potable water outlets and combination stop-and-waste valves shall not be installed underground or below grade. A freezeproof yard hydrant that drains the riser into the ground shall be considered as having a stop-and-waste valve below grade.

Exception: Freezeproof yard hydrants that drain the riser into the ground shall be permitted to be installed, provided that the potable water supply to such hydrants is protected in accordance with Section 608.14.2 or 608.14.5, and the hydrants and the piping from the backflow preventer to the hydrant are identified in accordance with Section 608.9.

608.9 Identification of nonpotable water systems. Where nonpotable water systems are installed, the piping conveying the nonpotable water shall be identified either by color marking, metal tags or tape in accordance with Sections 608.9.1 through 608.9.2.3.

608.9.1 Signage required. Nonpotable water outlets, such as hose connections, open ended pipes and faucets, shall be identified with signage that reads as follows: "Nonpotable water is utilized for [application name]. CAUTION: NONPOTABLE WATER – DO NOT DRINK." The words shall be legibly and indelibly printed on a tag or sign constructed of corrosion-resistant waterproof material or shall be indelibly printed on the fixture. The letters of the words shall be not less than 0.5 inch (12.7 mm) in height and in colors in contrast to the background on which they are applied. In addition to the required wordage, the pictograph shown in Figure 608.9.1 shall appear on the required signage.

**FIGURE 608.9.1
PICTOGRAPH—DO NOT DRINK**

608.9.2 Distribution pipe labeling and marking. Nonpotable distribution piping shall be purple in color and shall be embossed, or integrally stamped or marked, with the words: "CAUTION: NONPOTABLE WATER – DO NOT DRINK" or the piping shall be installed with a purple identification tape or wrap. Pipe identification shall include the contents of the piping system and an arrow indicating the direction of flow. Hazardous piping systems shall also contain information addressing the nature of the hazard. Pipe identification shall be repeated at intervals not exceeding 25 feet (7620 mm) and at each point where the piping passes through a wall, floor or roof. Lettering shall be readily observable within the room or space where the piping is located.

608.9.2.1 Color. The color of the pipe identification shall be discernable and consistent throughout the building. The color purple shall be used to identify reclaimed, rain and graywater distribution systems.

608.9.2.2 Lettering size. The size of the background color field and lettering shall comply with Table 608.9.2.2.

**TABLE 608.9.2.2
SIZE OF PIPE IDENTIFICATION**

PIPE DIAMETER (inches)	LENGTH BACKGROUND COLOR FIELD (inches)	SIZE OF LETTERS (inches)
3/4 to 1 1/4	8	0.5
1 1/2 to 2	8	0.75
2 1/2 to 6	12	1.25
8 to 10	24	2.5
Over 10	32	3.5

For SI   1 inch = 25.4 mm.

TABLE 608.1
APPLICATION OF BACKFLOW PREVENTERS

DEVICE	DEGREE OF HAZARD[a]	APPLICATION[b]	APPLICABLE STANDARDS
Backflow prevention assemblies:			
Double check backflow prevention assembly and double check fire protection backflow prevention assembly	Low hazard	Backpressure or backsiphonage Sizes $^3/_8''$–16″	ASSE 1015, AWWA C510, CSA B64.5, CSA B64.5.1
Double check detector fire protection backflow prevention assemblies	Low hazard	Backpressure or backsiphonage Sizes 2″–16″	ASSE 1048
Pressure vacuum breaker assembly	High or low hazard	Backsiphonage only Sizes $^1/_2''$–2″	ASSE 1020, CSA B64.1.2
Reduced pressure principle backflow prevention assembly and reduced pressure principle fire protection backflow assembly	High or low hazard	Backpressure or backsiphonage Sizes $^3/_8''$–16″	ASSE 1013, AWWA C511, CSA B64.4, CSA B64.4.1
Reduced pressure detector fire protection backflow prevention assemblies	High or low hazard	Backsiphonage or backpressure (Fire sprinkler systems)	ASSE 1047
Spill-resistant vacuum breaker assembly	High or low hazard	Backsiphonage only Sizes $^1/_4''$–2″	ASSE 1056; CSA B64.1.3
Backflow preventer plumbing devices:			
Antisiphon-type fill valves for gravity water closet flush tanks	High hazard	Backsiphonage only	ASSE 1002/ASME A112.1002/ CSA B125.12, CSA B125.3
Backflow preventer for carbonated beverage machines	Low hazard	Backpressure or backsiphonage Sizes $^1/_4''$–$^3/_8''$	ASSE 1022
Backflow preventer with intermediate atmospheric vents	Low hazard	Backpressure or backsiphonage Sizes $^1/_4''$–$^3/_4''$	ASSE 1012, CSA B64.3
Dual-check-valve-type backflow preventer	Low hazard	Backpressure or backsiphonage Sizes $^1/_4''$–1″	ASSE 1024, CSA B64.6
Hose connection backflow preventer	High or low hazard	Low head backpressure, rated working pressure, backpressure or backsiphonage Sizes $^1/_2''$–1″	ASME A112.21.3, ASSE 1052, CSA B64.2.1.1
Hose connection vacuum breaker	High or low hazard	Low head backpressure or backsiphonage Sizes $^1/_2''$, $^3/_4''$, 1″	ASME A112.21.3, ASSE 1011, CSA B64.2, CSA B64.2.1
Laboratory faucet backflow preventer	High or low hazard	Low head backpressure and backsiphonage	ASSE 1035, CSA B64.7
Pipe-applied atmospheric-type vacuum breaker	High or low hazard	Backsiphonage only Sizes $^1/_4''$–4″	ASSE 1001, CSA B64.1.1
Vacuum breaker wall hydrants, frost-resistant, automatic-draining type	High or low hazard	Low head backpressure or backsiphonage Sizes $^3/_4''$, 1″	ASME A112.21.3, ASSE 1019, CSA B64.2.2
Other means or methods:			
Air gap	High or low hazard	Backsiphonage or backpressure	ASME A112.1.2
Air gap fittings for use with plumbing fixtures, appliances and appurtenances	High or low hazard	Backsiphonage or backpressure	ASME A112.1.3
Barometric loop	High or low hazard	Backsiphonage only	(See Section 608.14.4)

For SI: 1 inch = 25.4 mm.

a. Low hazard—See Pollution (Section 202).
   High hazard—See Contamination (Section 202).
b. See Backpressure, low head (Section 202).
   See Backsiphonage (Section 202).

608.9.2.3 Identification tape. Where used, identification tape shall be not less than 3 inches (76 mm) wide and have white or black lettering on a purple field stating "CAUTION: NONPOTABLE WATER – DO NOT DRINK." Identification tape shall be installed on top of nonpotable rainwater distribution pipes, fastened not less than every 10 feet (3048 mm) to each pipe length and run continuously the entire length of the pipe.

608.10 Reutilization prohibited. Water utilized for the heating or cooling of equipment or other processes shall not be returned to the potable water system. Such water shall be discharged into a drainage system through an *air gap* or shall be utilized for nonpotable purposes.

608.11 Reuse of piping. Piping that has been utilized for any purpose other than conveying potable water shall not be utilized for conveying potable water.

608.12 Potable water tanks. Where in contact with potable water intended for drinking water, water tanks, coatings for the inside of tanks and liners for water tanks shall conform to NSF 61. The interior surface of a potable water tank shall not be lined, painted or repaired with any material that changes the taste, odor, color or potability of the water supply when the tank is placed in, or returned to, service.

608.13 Pumps and other appliances. Water pumps, filters, softeners, tanks and other devices that handle or treat potable water shall be protected against contamination.

608.14 Backflow protection. Means of protection against backflow shall be provided in accordance with Sections 608.14.1 through 608.14.9.

608.14.1 Air gap. The minimum required *air gap* shall be measured vertically from the lowest end of a potable water outlet to the *flood level rim* of the fixture or receptacle into which such potable water outlet discharges. *Air gaps* shall comply with ASME A112.1.2 and *air gap* fittings shall comply with ASME A112.1.3. Products that are listed and labeled to ASME A112.1.2 or ASME A112.1.3 shall be considered to be in compliance with this section.

608.14.2 Reduced pressure principle backflow prevention assemblies. Reduced pressure principle backflow prevention assemblies shall conform to ASSE 1013, AWWA C511, CSA B64.4 or CSA B64.4.1. Reduced pressure detector assembly backflow preventers shall conform to ASSE 1047. These devices shall be permitted to be installed where subject to continuous pressure conditions. The relief opening shall discharge by *air gap* and shall be prevented from being submerged.

608.14.3 Backflow preventer with intermediate atmospheric vent. Backflow preventers with intermediate atmospheric vents shall conform to ASSE 1012 or CSA B64.3. These devices shall be permitted to be installed where subject to continuous pressure conditions. The relief opening shall discharge by *air gap* and shall be prevented from being submerged.

608.14.4 Barometric loop. Barometric loops shall precede the point of connection and shall extend vertically to a height of 35 feet (10 668 mm). A barometric loop shall only be utilized as an atmospheric-type or pressure-type vacuum breaker.

608.14.5 Pressure vacuum breaker assemblies. Pressure vacuum breaker assemblies shall comply with ASSE 1020 or CSA B64.1.2. Spill-resistant vacuum breaker assemblies shall comply with ASSE 1056 or CSA B64.1.3. These assemblies shall be installed with the critical level of the assembly located not less than 12 inches (305 mm) above all downstream piping and outlets. Pressure vacuum breaker assemblies shall not be installed in locations where spillage could cause damage to the structure.

608.14.6 Atmospheric-type vacuum breakers. Pipe-applied atmospheric-type vacuum breakers shall conform to ASSE 1001 or CSA B64.1.1. Hose-connection vacuum breakers shall conform to ASME A112.21.3, ASSE 1011, ASSE 1019, ASSE 1035, ASSE 1052, CSA B64.2, CSA B64.2.1, CSA B64.2.1.1, CSA B64.2.2 or CSA B64.7. These devices shall operate under normal atmospheric pressure when the critical level is installed at the required height.

608.14.7 Double check backflow prevention assemblies. Double check backflow prevention assemblies shall conform to ASSE 1015, CSA B64.5, CSA B64.5.1 or AWWA C510. Double check detector fire protection backflow prevention assemblies shall conform to ASSE 1048. These assemblies shall be capable of operating under continuous pressure conditions.

608.14.8 Chemical dispenser backflow devices. Backflow devices for chemical dispensers shall comply with ASSE 1055 or shall be equipped with an *air gap* fitting.

608.14.9 Dual check backflow preventer. Dual check backflow preventers shall conform to ASSE 1024 or CSA B64.6.

608.15 Location of backflow preventers. Access shall be provided to backflow preventers as specified by the manufacturer's instructions.

608.15.1 Outdoor enclosures for backflow prevention devices. Outdoor enclosures for backflow prevention devices shall comply with ASSE 1060.

608.15.2 Protection of backflow preventers. Backflow preventers shall not be located in areas subject to freezing except where they can be removed by means of unions or are protected from freezing by heat, insulation or both.

608.15.2.1 Relief port piping. The termination of the piping from the relief port or *air gap* fitting of a backflow preventer shall discharge to an *approved* indirect waste receptor or to the outdoors where it will not cause damage or create a nuisance.

608.16 Protection of potable water outlets. Potable water openings and outlets shall be protected against backflow in accordance with Section 608.16.1, 608.16.2, 608.16.3, 608.16.4, 608.16.4.1 or 608.16.4.2.

608.16.1 Protection by air gap. Openings and outlets shall be protected by an *air gap* between the opening and the fixture *flood level rim* as specified in Table 608.16.1. Openings and outlets equipped for hose connection shall be protected by means other than an *air gap*.

TABLE 608.16.1
MINIMUM REQUIRED AIR GAPS

FIXTURE	MINIMUM AIR GAP	
	Away from a wall[a] (inches)	Close to a wall (inches)
Lavatories and other fixtures with effective openings not greater than $^1/_2$ inch in diameter	1	$1^1/_2$
Sinks, laundry trays, gooseneck back faucets and other fixtures with effective openings not greater than $^3/_4$ inch in diameter	$1^1/_2$	$2^1/_2$
Over-rim bath fillers and other fixtures with effective openings not greater than 1 inch in diameter	2	3
Drinking water fountains, single orifice not greater than $^7/_{16}$ inch in diameter or multiple orifices with a total area of 0.150 square inch (area of circle $^7/_{16}$ inch in diameter)	1	$1^1/_2$
Effective openings greater than 1 inch	Two times the diameter of the effective opening	Three times the diameter of the effective opening

For SI: 1 inch = 25.4 mm, 1 square inch = 645 mm^2.

a. Applicable where walls or obstructions are spaced from the nearest inside-edge of the spout opening a distance greater than three times the diameter of the effective opening for a single wall, or a distance greater than four times the diameter of the effective opening for two intersecting walls.

608.16.2 Protection by reduced pressure principle backflow prevention assembly. Openings and outlets shall be protected by a reduced pressure principle backflow prevention assembly or a reduced pressure principle fire protection backflow prevention assembly on potable water supplies.

608.16.3 Protection by a backflow preventer with intermediate atmospheric vent. Openings and outlets shall be protected by a backflow preventer with an intermediate atmospheric vent.

608.16.4 Protection by a vacuum breaker. Openings and outlets shall be protected by atmospheric-type or pressure-type vacuum breakers. The critical level of the vacuum breaker shall be set not less than 6 inches (152 mm) above the *flood level rim* of the fixture or device. Fill valves shall be set in accordance with Section 415.3.1. Vacuum breakers shall not be installed under exhaust hoods or similar locations that will contain toxic fumes or vapors. Pipe-applied vacuum breakers shall be installed not less than 6 inches (152 mm) above the *flood level rim* of the fixture, receptor or device served.

608.16.4.1 Deck-mounted and integral vacuum breakers. *Approved* deck-mounted or equipment-mounted vacuum breakers and faucets with integral atmospheric vacuum breakers or spill-resistant vacuum breaker assemblies shall be installed in accordance with the manufacturer's instructions and the requirements for labeling with the critical level not less than 1 inch (25 mm) above the *flood level rim*.

608.16.4.2 Hose connections. Sillcocks, hose bibbs, wall hydrants and other openings with a hose connection shall be protected by an atmospheric-type or pressure-type vacuum breaker or a permanently attached hose connection vacuum breaker.

Exceptions:

1. This section shall not apply to water heater and boiler drain valves that are provided with hose connection threads and that are intended only for tank or vessel draining.

2. This section shall not apply to water supply valves intended for connection of clothes washing machines where backflow prevention is otherwise provided or is integral with the machine.

608.17 Connections to the potable water system. Connections to the potable water system shall conform to Sections 608.17.1 through 608.17.10.

608.17.1 Beverage dispensers. The water supply connection to beverage dispensers shall be protected against backflow in accordance with Sections 608.17.1.1 and 608.17.1.2.

608.17.1.1 Carbonated beverage dispensers. The water supply connection to each carbonated beverage dispenser shall be protected against backflow by a backflow preventer conforming to ASSE 1022 or by an *air gap*. The portion of the backflow preventer device downstream from the second check valve of the device and the piping downstream therefrom shall not be affected by carbon dioxide gas.

608.17.1.2 Coffee machines and noncarbonated drink dispensers. The water supply connection to each coffee machine and each noncarbonated beverage dispenser shall be protected against backflow by a backflow preventer conforming to ASSE 1022 or ASSE 1024, or protected by an *air gap*.

608.17.2 Connections to boilers. The potable supply to the boiler shall be equipped with a backflow preventer with an intermediate atmospheric vent complying with ASSE 1012 or CSA B64.3. Where conditioning chemicals are introduced into the system, the potable water connection shall be protected by an *air gap* or a reduced pressure principle backflow preventer, complying with ASSE 1013, CSA B64.4 or AWWA C511.

608.17.3 Heat exchangers. Heat exchangers utilizing an essentially toxic transfer fluid shall be separated from the potable water by double-wall construction. An *air gap* open to the atmosphere shall be provided between the two walls. Heat exchangers utilizing an essentially nontoxic transfer fluid shall be permitted to be of single-wall construction.

608.17.4 Connections to automatic fire sprinkler systems and standpipe systems. The potable water supply to automatic fire sprinkler and standpipe systems shall be protected against backflow by a double check backflow prevention assembly, a double check fire protection backflow prevention assembly or a reduced pressure principle fire protection backflow prevention assembly.

Exceptions:

1. Where systems are installed as a portion of the water distribution system in accordance with the requirements of this code and are not provided with a fire department connection, isolation of the water supply system shall not be required.

2. Isolation of the water distribution system is not required for deluge, preaction or dry pipe systems.

608.17.4.1 Additives or nonpotable source. Where systems under continuous pressure contain chemical additives or antifreeze, or where systems are connected to a nonpotable secondary water supply, the potable water supply shall be protected against backflow by a reduced pressure principle backflow prevention assembly or a reduced pressure principle fire protection backflow prevention assembly. Where chemical additives or antifreeze are added to only a portion of an automatic fire sprinkler or standpipe system, the reduced pressure principle backflow prevention assembly or the reduced pressure principle fire protection backflow prevention assembly shall be permitted to be located so as to isolate that portion of the system. Where systems are not under continuous pressure, the potable water supply shall be protected against backflow by an air gap or an atmospheric vacuum breaker conforming to ASSE 1001 or CSA B64.1.1.

608.17.5 Connections to lawn irrigation systems. The potable water supply to lawn irrigation systems shall be protected against backflow by an atmospheric vacuum breaker, a pressure vacuum breaker assembly or a reduced pressure principle backflow prevention assembly. Valves shall not be installed downstream from an atmospheric vacuum breaker. Where chemicals are introduced into the system, the potable water supply shall be protected against backflow by a reduced pressure principle backflow prevention assembly.

608.17.6 Connections subject to backpressure. Where a potable water connection is made to a nonpotable line, fixture, tank, vat, pump or other equipment subject to high-hazard backpressure, the potable water connection shall be protected by a reduced pressure principle backflow prevention assembly.

608.17.7 Chemical dispensers. Where chemical dispensers connect to the potable water distribution system, the water supply system shall be protected against backflow in accordance with Section 608.14.1, 608.14.2, 608.14.5, 608.14.6 or 608.14.8.

608.17.8 Portable cleaning equipment. Where the portable cleaning equipment connects to the water distribution system, the water supply system shall be protected against backflow in accordance with Section 608.14.1, 608.14.2, 608.14.3, 608.14.7 or 608.14.8.

608.17.9 Dental pumping equipment. The water supply connection to each dental pumping equipment system, the water supply system shall be protected against backflow in accordance with Section 608.14.1, 608.14.2, 608.14.5, 608.14.6 or 608.14.8.

608.17.10 Humidifiers. The water supply connection to humidifiers that do not have internal backflow protection shall be protected against backflow by a backflow preventer conforming to ASSE 1012 or by an *air gap*.

608.18 Protection of individual water supplies. An individual water supply shall be located and constructed so as to be safeguarded against contamination in accordance with Sections 608.18.1 through 608.18.8.

608.18.1 Well locations. A potable ground water source or pump suction line shall not be located closer to potential sources of contamination than the distances shown in Table 608.18.1. In the event the underlying rock structure is limestone or fragmented shale, the local or state health department shall be consulted on well site location. The distances in Table 608.18.1 constitute minimum separation and shall be increased in areas of creviced rock or limestone, or where the direction of movement of the ground water is from sources of contamination toward the well.

TABLE 608.18.1
DISTANCE FROM CONTAMINATION TO PRIVATE WATER SUPPLIES AND PUMP SUCTION LINES

SOURCE OF CONTAMINATION	DISTANCE (feet)
Barnyard	100
Farm silo	25
Pasture	100
Pumphouse floor drain of cast iron draining to ground surface	2
Seepage pits	50
Septic tank	25
Sewer	10
Subsurface disposal fields	50
Subsurface pits	50

For SI: 1 foot = 304.8 mm.

608.18.2 Elevation. Well sites shall be positively drained and shall be at higher elevations than potential sources of contamination.

608.18.3 Depth. Private potable well supplies shall not be developed from a water table less than 10 feet (3048 mm) below the ground surface.

608.18.4 Water-tight casings. Each well shall be provided with a water-tight casing extending to not less than 10 feet (3048 mm) below the ground surface. Casings shall extend not less than 6 inches (152 mm) above the well platform. Casings shall be large enough to permit installation of a separate drop pipe. Casings shall be sealed at the bottom in an impermeable stratum or extend several feet into the water-bearing stratum.

608.18.5 Drilled or driven well casings. Drilled or driven well casings shall be of steel or other *approved* material. Where drilled wells extend into a rock formation, the well casing shall extend to and set firmly in the formation. The annular space between the earth and the outside of the casing shall be filled with cement grout to a depth of not less than 10 feet (3048 mm) below the ground surface. In an instance of casing to rock installation, the grout shall extend to the rock surface.

608.18.6 Dug or bored well casings. Dug or bored well casings shall be of water-tight concrete, tile or galvanized or corrugated metal pipe extending to not less than 10 feet (3048 mm) below the ground surface. Where the water table is more than 10 feet (3048 mm) below the ground surface, the water-tight casing shall extend below the table surface. Well casings for dug wells or bored wells constructed with sections of concrete, tile or galvanized or corrugated metal pipe shall be surrounded by 6 inches (152 mm) of grout poured into the hole between the outside of the casing and the ground and extending not less than 10 feet (3048 mm) below the ground surface.

608.18.7 Cover. Potable water wells shall be equipped with an overlapping water-tight cover at the top of the well casing or pipe sleeve such that contaminated water or other substances are prevented from entering the well through the annular opening at the top of the well casing, wall or pipe sleeve. Covers shall extend downward not less than 2 inches (51 mm) over the outside of the well casing or wall. A dug well cover shall be provided with a pipe sleeve permitting the withdrawal of the pump suction pipe, cylinder or jet body without disturbing the cover. Where pump sections or discharge pipes enter or leave a well through the side of the casing, the circle of contact shall be watertight.

608.18.8 Drainage. Potable water wells and springs shall be constructed such that surface drainage will be diverted away from the well or spring.

SECTION 609
HEALTH CARE PLUMBING

609.1 Scope. This section shall govern those aspects of health care plumbing systems that differ from plumbing systems in other structures. Health care plumbing systems shall conform to the requirements of this section in addition to the other requirements of this code. The provisions of this section shall apply to the special devices and equipment installed and maintained in the following occupancies: Group I-1, Group I-2, Group B ambulatory care facilities, medical offices, research and testing laboratories, and Group F facilities manufacturing pharmaceutical drugs and medicines.

609.2 Water service. Hospitals shall have two water service pipes installed in such a manner so as to minimize the potential for an interruption of the supply of water in the event of a water main or water service pipe failure.

609.3 Hot water. *Hot water* shall be provided to supply all of the hospital fixture, kitchen and laundry requirements. Special fixtures and equipment shall have hot water supplied at a temperature specified by the manufacturer. The hot water system shall be installed in accordance with Section 607.

609.4 Vacuum breaker installation. Vacuum breakers shall be installed not less than 6 inches (152 mm) above the *flood level rim* of the fixture or device in accordance with Section 608. The *flood level rim* of hose connections shall be the maximum height at which any hose is utilized.

609.5 Prohibited water closet and clinical sink supply. Jet- or water-supplied orifices, except those supplied by the flush connections, shall not be located in or connected with a water closet bowl or clinical sink. This section shall not prohibit an *approved* bidet installation.

609.6 Clinical, hydrotherapeutic and radiological equipment. Clinical, hydrotherapeutic, radiological or any equipment that is supplied with water or that discharges to the waste system shall conform to the requirements of this section and Section 608.

609.7 Condensate drain trap seal. A water supply shall be provided for cleaning, flushing and resealing the condensate trap, and the trap shall discharge through an *air gap* in accordance with Section 608.

609.8 Valve leakage diverter. Each water sterilizer filled with water through directly connected piping shall be equipped with an *approved* leakage diverter or bleed line on the water supply control valve to indicate and conduct any leakage of unsterile water away from the sterile zone.

SECTION 610
DISINFECTION OF POTABLE WATER SYSTEM

610.1 General. New potable water systems shall be purged of deleterious matter and disinfected prior to utilization. The method to be followed shall be that prescribed by the health authority or water purveyor having jurisdiction or, in the absence of a prescribed method, the procedure described in either AWWA C651 or AWWA C652, or as described in this section. This requirement shall apply to "on-site" or "in-plant" fabrication of a system or to a modular portion of a system.

1. The pipe system shall be flushed with clean, potable water until dirty water does not appear at the points of outlet.

2. The system or part thereof shall be filled with a water/chlorine solution containing not less than 50 parts per million (50 mg/L) of chlorine, and the system or part thereof shall be valved off and allowed to stand for 24 hours; or the system or part thereof shall be filled with a water/chlorine solution containing not less than 200 parts per million (200 mg/L) of chlorine and allowed to stand for 3 hours.

3. Following the required standing time, the system shall be flushed with clean potable water until the chlorine is purged from the system.

4. The procedure shall be repeated where shown by a bacteriological examination that contamination remains present in the system.

SECTION 611
DRINKING WATER TREATMENT UNITS

611.1 Design. Point-of-use reverse osmosis drinking water treatment units shall comply with NSF 58 or CSA B483.1. Drinking water treatment units shall meet the requirements of NSF 42, NSF 44, NSF 53, NSF 62 or CSA B483.1.

611.2 Reverse osmosis systems. The discharge from a reverse osmosis drinking water treatment unit shall enter the drainage system through an *air gap* or an *air gap* device that meets the requirements of NSF 58 or CSA B483.1.

611.3 Connection tubing. The tubing to and from drinking water treatment units shall be of a size and material as recommended by the manufacturer. The tubing shall comply with NSF 14, NSF 42, NSF 44, NSF 53, NSF 58 or NSF 61.

SECTION 612
SOLAR SYSTEMS

612.1 Solar systems. The construction, installation, alterations and repair of systems, equipment and appliances intended to utilize solar energy for space heating or cooling, domestic hot water heating, swimming pool heating or process heating shall be in accordance with the *International Mechanical Code*.

SECTION 613
TEMPERATURE CONTROL DEVICES AND VALVES

613.1 Temperature-actuated mixing valves. Temperature-actuated mixing valves, which are installed to reduce water temperatures to defined limits, shall comply with ASSE 1017. Such valves shall be installed at the hot water source.

CHAPTER 7

SANITARY DRAINAGE

User note:

About this chapter: Chapter 7 regulates the methods and piping systems that remove water that has served a purpose such as flushing water closets, bathing, culinary activities and equipment discharges. The types of materials, drainage fitting and the connection methods are covered for these systems that begin at the receiving fixtures and end at the point of disposal for the liquid waste. A design method for a gravity flow system of vertical and horizontal piping is provided based on the probability of flows from specific fixtures. Vacuum and pumped types of liquid waste removal methods are also regulated by this chapter.

SECTION 701
GENERAL

701.1 Scope. The provisions of this chapter shall govern the materials, design, construction and installation of sanitary drainage systems.

701.2 Connection to sewer required. Sanitary drainage piping from plumbing fixtures in buildings and sanitary drainage piping systems from premises shall be connected to a public *sewer*. Where a public *sewer* is not available, the sanitary drainage piping and systems shall be connected to a private sewage disposal system in compliance with state or local requirements. Where state or local requirements do not exist for private sewage disposal systems, the sanitary drainage piping and systems shall be connected to an approved private sewage disposal system that is in accordance with the *International Private Sewage Disposal Code*.

> **Exception:** Sanitary drainage piping and systems that convey only the discharge from bathtubs, showers, lavatories, clothes washers and laundry trays shall not be required to connect to a public *sewer* or to a private sewage disposal system provided that the piping or systems are connected to a system in accordance with Chapter 13 or 14.

701.3 Separate sewer connection. A building having plumbing fixtures installed and intended for human habitation, occupancy or use on premises abutting on a street, alley or easement in which there is a public *sewer* shall have a separate connection with the *sewer*. Where located on the same lot, multiple buildings shall not be prohibited from connecting to a common *building sewer* that connects to the public *sewer*.

701.4 Sewage treatment. Sewage or other waste from a plumbing system that is deleterious to surface or subsurface waters shall not be discharged into the ground or into any waterway unless it has first been rendered innocuous through subjection to an *approved* form of treatment.

701.5 Damage to drainage system or public sewer. Waste detrimental to the public *sewer* system or to the functioning of the sewage-treatment plant shall be treated and disposed of in accordance with Section 1003 as directed by the code official.

701.6 Tests. The sanitary drainage system shall be tested in accordance with Section 312.

701.7 Engineered systems. Engineered sanitary drainage systems shall conform to the provisions of Sections 316 and 713.

SECTION 702
MATERIALS

702.1 Above-ground sanitary drainage and vent pipe. Above-ground soil, waste and vent pipe shall conform to one of the standards listed in Table 702.1.

TABLE 702.1
ABOVE-GROUND DRAINAGE AND VENT PIPE

MATERIAL	STANDARD
Acrylonitrile butadiene styrene (ABS) plastic pipe in IPS diameters, including Schedule 40, DR 22 (PS 200) and DR 24 (PS 140); with a solid, cellular core or composite wall	ASTM D2661; ASTM F628; ASTM F1488; CSA B181.1
Cast-iron pipe	ASTM A74; ASTM A888; CISPI 301
Copper or copper-alloy pipe	ASTM B42; ASTM B43; ASTM B302
Copper or copper-alloy tubing (Type K, L, M or DWV)	ASTM B75; ASTM B88; ASTM B251; ASTM B306
Galvanized steel pipe	ASTM A53
Glass pipe	ASTM C1053
Polyolefin pipe	ASTM F1412; CSA B181.3
Polyvinyl chloride (PVC) plastic pipe in IPS diameters, including Schedule 40, DR 22 (PS 200), and DR 24 (PS 140); with a solid, cellular core or composite wall	ASTM D2665; ASTM F891; ASTM F1488; CSA B181.2
Polyvinyl chloride (PVC) plastic pipe with a 3.25-inch O.D. and a solid, cellular core or composite wall	ASTM D2949, ASTM F1488
Polyvinylidene fluoride (PVDF) plastic pipe	ASTM F1673; CSA B181.3
Stainless steel drainage systems, Types 304 and 316L	ASME A112.3.1

2018 INTERNATIONAL PLUMBING CODE®

SANITARY DRAINAGE

702.2 Underground building sanitary drainage and vent pipe. Underground building sanitary drainage and vent pipe shall conform to one of the standards listed in Table 702.2.

TABLE 702.2
UNDERGROUND BUILDING DRAINAGE AND VENT PIPE

MATERIAL	STANDARD
Acrylonitrile butadiene styrene (ABS) plastic pipe in IPS diameters, including Schedule 40, DR 22 (PS 200) and DR 24 (PS 140); with a solid, cellular core or composite wall	ASTM D2661; ASTM F628; ASTM F1488; CSA B181.1
Cast-iron pipe	ASTM A74; ASTM A888; CISPI 301
Copper or copper-alloy tubing (Type K, L, M or DWV)	ASTM B75; ASTM B88; ASTM B251; ASTM B306
Polyethylene (PE) plastic pipe (SDR-PR)	ASTM F714
Polyolefin pipe	ASTM F1412; ASTM F714; CSA B181.3
Polyvinyl chloride (PVC) plastic pipe in IPS diameters, including Schedule 40, DR 22 (PS 200) and DR 24 (PS 140); with a solid, cellular core or composite wall	ASTM D2665; ASTM F891; ASTM F1488; CSA B181.2
Polyvinyl chloride (PVC) plastic pipe with a 3.25-inch O.D. and a solid, cellular core or composite wall	ASTM D2949, ASTM F1488
Polyvinylidene fluoride (PVDF) plastic pipe	ASTM F1673; CSA B181.3
Stainless steel drainage systems, Type 316L	ASME A112.3.1

For SI: 1 inch = 25.4 mm.

702.3 Building sewer pipe. *Building sewer* pipe shall conform to one of the standards listed in Table 702.3.

702.4 Fittings. Pipe fittings shall be *approved* for installation with the piping material installed and shall comply with the applicable standards listed in Table 702.4.

702.5 Temperature rating. Where the wastewater temperature will be greater than 140°F (60°C), the sanitary drainage piping material shall be rated for the highest temperature of the wastewater.

702.6 Chemical waste system. A chemical waste system shall be completely separated from the sanitary drainage system. The chemical waste shall be treated in accordance with Section 803.2 before discharging to the sanitary drainage system. Separate drainage systems for chemical wastes and vent pipes shall be of an *approved* material that is resistant to corrosion and degradation for the concentrations of chemicals involved.

702.7 Lead bends and traps. The wall thickness of lead bends and traps shall be not less than $^1/_8$ inch (3.2 mm).

TABLE 702.3
BUILDING SEWER PIPE

MATERIAL	STANDARD
Acrylonitrile butadiene styrene (ABS) plastic pipe in IPS diameters, including Schedule 40, DR 22 (PS 200) and DR 24 (PS 140); with a solid, cellular core or composite wall	ASTM D2661; ASTM F628; ASTM F1488; CSA B181.1
Acrylonitrile butadiene styrene (ABS) plastic pipe in sewer and drain diameters, including SDR 42 (PS 20), PS 35, SDR 35 (PS 45), PS 50, PS 100, PS 140, SDR 23.5 (PS 150) and PS 200; with a solid, cellular core or composite wall	ASTM F1488; ASTM D2751
Cast-iron pipe	ASTM A74; ASTM A888; CISPI 301
Concrete pipe	ASTM C14; ASTM C76; CSA A257.1M; CSA A257.2M
Copper or copper-alloy tubing (Type K or L)	ASTM B75; ASTM B88; ASTM B251
Polyethylene (PE) plastic pipe (SDR-PR)	ASTM F714
Polypropylene (PP) plastic pipe	ASTM F2736; ASTM F2764; CSA B182.13
Polyvinyl chloride (PVC) plastic pipe in IPS diameters, including Schedule 40, DR 22 (PS 200) and DR 24 (PS 140); with a solid, cellular core or composite wall	ASTM D2665; ASTM F891; ASTM F1488
Polyvinyl chloride (PVC) plastic pipe in sewer and drain diameters, including PS 25, SDR 41 (PS 28), PS 35, SDR 35 (PS 46), PS 50, PS 100, SDR 26 (PS 115), PS 140 and PS 200; with a solid, cellular core or composite wall	ASTM F891; ASTM F1488; ASTM D3034; CSA B182.2; CSA B182.4
Polyvinyl chloride (PVC) plastic pipe with a 3.25-inch O.D. and a solid, cellular core or composite wall	ASTM D2949, ASTM F1488
Polyvinylidene fluoride (PVDF) plastic pipe	ASTM F1673; CSA B181.3
Stainless steel drainage systems, Types 304 and 316L	ASME A112.3.1
Vitrified clay pipe	ASTM C4; ASTM C700

For SI: 1 inch = 25.4 mm.

SANITARY DRAINAGE

TABLE 702.4
PIPE FITTINGS

MATERIAL	STANDARD
Acrylonitrile butadiene styrene (ABS) plastic pipe in IPS diameters	ASTM D2661; ASTM F628; CSA B181.1
Acrylonotrile butadiene styrene (ABS) plastic pipe in sewer and drain diameters	ASTM D2751
Cast iron	ASME B16.4; ASME B16.12; ASTM A74; ASTM A888; CISPI 301
Copper or copper alloy	ASME B16.15; ASME B16.18; ASME B16.22; ASME B16.23; ASME B16.26; ASME B16.29
Glass	ASTM C1053
Gray iron and ductile iron	AWWA C110/A21.10
Polyethylene	ASTM D2683
Polyolefin	ASTM F1412; CSA B181.3
Polyvinyl chloride (PVC) plastic in IPS diameters	ASTM D2665; ASTM F1866
Polyvinyl chloride (PVC) plastic pipe in sewer and drain diameters	ASTM D3034
Polyvinyl chloride (PVC) plastic pipe with a 3.25-inch O.D.	ASTM D2949
Polyvinylidene fluoride (PVDF) plastic pipe	ASTM F1673; CSA B181.3
Stainless steel drainage systems, Types 304 and 316L	ASME A112.3.1
Steel	ASME B16.9; ASME B16.11; ASME B16.28
Vitrified clay	ASTM C700

For SI: 1 inch = 25.4 mm.

SECTION 703
BUILDING SEWER

703.1 Building sewer pipe near the water service. The proximity of a *sewer* to a water service shall comply with Section 603.2.

703.2 Drainage pipe in filled ground. Where a *building sewer* or *building drain* is installed on filled or unstable ground, the drainage pipe shall conform to one of the standards for ABS plastic pipe, cast-iron pipe, copper or copper-alloy tubing, PVC plastic pipe or polypropylene plastic pipe indicated in Table 702.3.

703.3 Sanitary and storm sewers. Where separate systems of sanitary drainage and storm drainage are installed in the same property, the sanitary and storm building sewers or drains shall be permitted to be laid side by side in one trench.

703.4 Existing building sewers and building drains. Where the entire sanitary drainage system of an existing building is replaced, existing *building drains* under concrete slabs and existing building sewers that will serve the new system shall be internally examined to verify that the piping is sloping in the correct direction, is not broken, is not obstructed and is sized for the drainage load of the new plumbing drainage system to be installed.

703.5 Cleanouts on building sewers. Cleanouts on *building sewers* shall be located as indicated in Section 708.

703.6 Combined sanitary and storm public sewer. Where the public sewer is a combined system for both sanitary and storm water, the sanitary sewer shall be connected independently to the public sewer.

SECTION 704
DRAINAGE PIPING INSTALLATION

704.1 Slope of horizontal drainage piping. Horizontal drainage piping shall be installed in uniform alignment at uniform slopes. The slope of a horizontal drainage pipe shall be not less than that indicated in Table 704.1 except that where the drainage piping is upstream of a grease interceptor, the slope of the piping shall be not less than $1/4$ inch per foot (2-percent slope).

TABLE 704.1
SLOPE OF HORIZONTAL DRAINAGE PIPE

SIZE (inches)	MINIMUM SLOPE (inch per foot)
$2^1/_2$ or less	$1/_4$ [a]
3 to 6	$1/_8$ [a]
8 or larger	$1/_{16}$ [a]

For SI: 1 inch = 25.4 mm, 1 inch per foot = 83.33 mm/m.

a. Slopes for piping draining to a grease interceptor shall comply with Section 704.1.

704.2 Reduction in pipe size in the direction of flow. The size of the drainage piping shall not be reduced in the direction of the flow. The following shall not be considered as a reduction in size in the direction of flow:

1. A 4-inch by 3-inch (102 mm by 76 mm) water closet flange.

2. A water closet bend fitting having a 4-inch (102 mm) inlet and a 3-inch (76 mm) outlet provided that the 4-inch leg of the fitting is upright and below, but not necessarily directly connected to, the water closet flange.

3. An offset closet flange.

704.3 Connections to offsets and bases of stacks. Horizontal *branches* shall connect to the bases of stacks at a point located not less than 10 times the diameter of the drainage *stack* downstream from the *stack*. Horizontal *branches* shall connect to horizontal *stack* offsets at a point located not less than 10 times the diameter of the drainage *stack* downstream from the upper *stack*.

704.4 Future fixtures. Drainage piping for future fixtures shall terminate with an *approved* cap or plug.

SANITARY DRAINAGE

SECTION 705
JOINTS

705.1 General. This section contains provisions applicable to joints specific to sanitary drainage piping.

705.2 ABS plastic. Joints between ABS plastic pipe or fittings shall comply with Sections 705.2.1 through 705.2.3.

705.2.1 Mechanical joints. Mechanical joints on drainage pipes shall be made with an elastomeric seal conforming to ASTM C1173, ASTM D3212 or CSA B602. Mechanical joints shall be installed only in underground systems unless otherwise *approved*. Joints shall be installed in accordance with the manufacturer's instructions.

705.2.2 Solvent cementing. Joint surfaces shall be clean and free from moisture. Solvent cement that conforms to ASTM D2235 or CSA B181.1 shall be applied to all joint surfaces. The joint shall be made while the cement is wet. Joints shall be made in accordance with ASTM D2235, ASTM D2661, ASTM F628 or CSA B181.1. Solvent-cement joints shall be permitted above or below ground.

705.2.3 Threaded joints. Threads shall conform to ASME B1.20.1. Schedule 80 or heavier pipe shall be permitted to be threaded with dies specifically designed for plastic pipe. *Approved* thread lubricant or tape shall be applied on the male threads only.

705.3 Cast iron. Joints between cast-iron pipe or fittings shall comply with Sections 705.3.1 through 705.3.3.

705.3.1 Caulked joints. Joints for hub and spigot pipe shall be firmly packed with oakum or hemp. Molten lead shall be poured in one operation to a depth of not less than 1 inch (25 mm). The lead shall not recede more than $^1/_8$ inch (3.2 mm) below the rim of the hub and shall be caulked tight. Paint, varnish or other coatings shall not be permitted on the jointing material until after the joint has been tested and *approved*. Lead shall be run in one pouring and shall be caulked tight. Acid-resistant rope and acidproof cement shall be permitted.

705.3.2 Compression gasket joints. Compression gaskets for hub and spigot pipe and fittings shall conform to ASTM C564 and shall be tested to ASTM C1563. Gaskets shall be compressed when the pipe is fully inserted.

705.3.3 Mechanical joint coupling. Mechanical joint couplings for hubless pipe and fittings shall consist of an elastomeric sealing sleeve and a metallic shield that comply with CISPI 310, ASTM C1277 or ASTM C1540. The elastomeric sealing sleeve shall conform to ASTM C564 or CSA B602 and shall be provided with a center stop. Mechanical joint couplings shall be installed in accordance with the manufacturer's instructions.

705.4 Concrete joints. Joints between concrete pipe and fittings shall be made with an elastomeric seal conforming to ASTM C443, ASTM C1173, CSA A257.3M or CSA B602.

705.5 Copper pipe. Joints between copper or copper-alloy pipe or fittings shall comply with Sections 705.5.1 through 705.5.5.

705.5.1 Brazed joints. Joint surfaces shall be cleaned. An *approved* flux shall be applied where required. The joint shall be brazed with a filler metal conforming to AWS A5.8.

705.5.2 Mechanical joints. Mechanical joints shall be installed in accordance with the manufacturer's instructions.

705.5.3 Solder joints. Solder joints shall be made in accordance with the methods of ASTM B828. Cut tube ends shall be reamed to the full inside diameter of the tube end. Joint surfaces shall be cleaned. A flux conforming to ASTM B813 shall be applied. The joint shall be soldered with a solder conforming to ASTM B32.

705.5.4 Threaded joints. Threads shall conform to ASME B1.20.1. Pipe-joint compound or tape shall be applied on the male threads only.

705.5.5 Welded joints. Joint surfaces shall be cleaned. The joint shall be welded with an *approved* filler metal.

705.6 Copper tubing. Joints between copper or copper-alloy tubing or fittings shall comply with Sections 705.6.1 through 705.6.3.

705.6.1 Brazed joints. Joint surfaces shall be cleaned. An *approved* flux shall be applied where required. The joint shall be brazed with a filler metal conforming to AWS A5.8.

705.6.2 Mechanical joints. Mechanical joints shall be installed in accordance with the manufacturer's instructions.

705.6.3 Solder joints. Solder joints shall be made in accordance with the methods of ASTM B828. Cut tube ends shall be reamed to the full inside diameter of the tube end. Joint surfaces shall be cleaned. A flux conforming to ASTM B813 shall be applied. The joint shall be soldered with a solder conforming to ASTM B32.

705.7 Borosilicate glass joints. Glass-to-glass connections shall be made with a bolted compression-type, 300 series stainless steel coupling with contoured acid-resistant elastomeric compression ring and a fluorocarbon polymer inner seal ring; or with caulked joints in accordance with Section 705.7.1.

705.7.1 Caulked joints. Lead-caulked joints for hub and spigot soil pipe shall be firmly packed with oakum or hemp and filled with molten lead not less than 1 inch (25 mm) in depth and not to recede more than $^1/_8$ inch (3.2 mm) below the rim of the hub. Paint, varnish or other coatings shall not be permitted on the jointing material until after the joint has been tested and *approved*. Lead shall be run in one pouring and shall be caulked tight. Acid-resistant rope and acidproof cement shall be permitted.

705.8 Steel. Joints between galvanized steel pipe or fittings shall comply with Sections 705.8.1 and 705.8.2.

705.8.1 Threaded joints. Threads shall conform to ASME B1.20.1. Pipe-joint compound or tape shall be applied on the male threads only.

705.8.2 Mechanical joints. Joints shall be made with an *approved* elastomeric seal. Mechanical joints shall be installed in accordance with the manufacturer's instructions.

705.9 Lead. Joints between lead pipe or fittings shall comply with Sections 705.9.1 and 705.9.2.

705.9.1 Burned. Burned joints shall be uniformly fused together into one continuous piece. The thickness of the joint shall be not less than the thickness of the lead being joined. The filler metal shall be of the same material as the pipe.

705.9.2 Wiped. Joints shall be fully wiped, with an exposed surface on each side of the joint not less than $^3/_4$ inch (19.1 mm). The joint shall be not less than $^3/_8$ inch (9.5 mm) thick at the thickest point.

705.10 PVC plastic. Joints between PVC plastic pipe or fittings shall comply with Sections 705.10.1 through 705.10.3.

705.10.1 Mechanical joints. Mechanical joints on drainage pipe shall be made with an elastomeric seal conforming to ASTM C1173, ASTM D3212 or CSA B602. Mechanical joints shall not be installed in above-ground systems, unless otherwise *approved*. Joints shall be installed in accordance with the manufacturer's instructions.

705.10.2 Solvent cementing. Joint surfaces shall be clean and free from moisture. A purple primer that conforms to ASTM F656 shall be applied. Solvent cement not purple in color and conforming to ASTM D2564, CSA B137.3, CSA B181.2 or CSA B182.1 shall be applied to all joint surfaces. The joint shall be made while the cement is wet and shall be in accordance with ASTM D2855. Solvent-cement joints shall be permitted above or below ground.

> **Exception:** A primer is not required where both of the following conditions apply:
> 1. The solvent cement used is third-party certified as conforming to ASTM D2564.
> 2. The solvent cement is used only for joining PVC drain, waste and vent pipe and fittings in non-pressure applications in sizes up to and including 4 inches (102 mm) in diameter.

705.10.3 Threaded joints. Threads shall conform to ASME B1.20.1. Where pipe is to be threaded, the pipe shall have a wall thickness of not less than Schedule 80. Pipe threads shall be made with dies specifically designed for plastic pipe. *Approved* thread lubricant or tape shall be applied on the male threads only.

705.11 Vitrified clay. Joints between vitrified clay pipe or fittings shall be made with an elastomeric seal conforming to ASTM C425, ASTM C1173 or CSA B602.

705.12 Polyethylene plastic pipe. Joints between polyethylene plastic pipe and fittings shall be underground and shall comply with Section 705.12.1 or 705.12.2.

705.12.1 Heat-fusion joints. Joint surfaces shall be clean and free from moisture. Joint surfaces shall be cut, heated to melting temperature and joined using tools specifically designed for the operation. Joints shall be undisturbed until cool. Joints shall be made in accordance with ASTM D2657 and the manufacturer's instructions.

705.12.2 Mechanical joints. Mechanical joints in drainage piping shall be made with an elastomeric seal conforming to ASTM C1173, ASTM D3212 or CSA B602. Mechanical joints shall be installed in accordance with the manufacturer's instructions.

705.13 Polyolefin plastic. Joints between polyolefin plastic pipe and fittings shall comply with Sections 705.13.1 and 705.13.2.

705.13.1 Heat-fusion joints. Heat-fusion joints for polyolefin pipe and tubing joints shall be installed with socket-type heat-fused polyolefin fittings or electrofusion polyolefin fittings. Joint surfaces shall be clean and free from moisture. The joint shall be undisturbed until cool. Joints shall be made in accordance with ASTM F1412 or CSA B181.3.

705.13.2 Mechanical and compression sleeve joints. Mechanical and compression sleeve joints shall be installed in accordance with the manufacturer's instructions.

705.14 Polyvinylidene fluoride plastic. Joints between polyvinylidene plastic pipe and fittings shall comply with Sections 705.14.1 and 705.14.2.

705.14.1 Heat-fusion joints. Heat-fusion joints for polyvinylidene fluoride pipe and tubing joints shall be installed with socket-type heat-fused polyvinylidene fluoride fittings or electrofusion polyvinylidene fittings and couplings. Joint surfaces shall be clean and free from moisture. The joint shall be undisturbed until cool. Joints shall be made in accordance with ASTM F1673.

705.14.2 Mechanical and compression sleeve joints. Mechanical and compression sleeve joints shall be installed in accordance with the manufacturer's instructions.

705.15 Polypropylene plastic. The joint between polypropylene plastic pipe and fittings shall incorporate an elastomeric seal. The joint shall conform to ASTM D3212. Mechanical joints shall not be installed above ground.

705.16 Joints between different materials. Joints between different piping materials shall be made with a mechanical joint of the compression or mechanical-sealing type conforming to ASTM C1173, ASTM C1460 or ASTM C1461. Connectors and adapters shall be *approved* for the application and such joints shall have an elastomeric seal conforming to ASTM C425, ASTM C443, ASTM C564, ASTM C1440, ASTM F477, CSA A257.3M or CSA B602, or as required in Sections 705.16.1 through 705.16.7. Joints between glass pipe and other types of materials shall be made with adapters having a TFE seal. Joints shall be installed in accordance with the manufacturer's instructions.

705.16.1 Copper pipe or tubing to cast-iron hub pipe. Joints between copper pipe or tubing and cast-iron hub pipe shall be made with a copper or copper alloy ferrule or compression joint. The copper pipe or tubing shall be soldered to the ferrule in an *approved* manner, and the ferrule shall be joined to the cast-iron hub by a caulked joint or a mechanical compression joint.

SANITARY DRAINAGE

705.16.2 Copper or copper-alloy pipe or tubing to galvanized steel pipe. Joints between copper or copper-alloy pipe or tubing and galvanized steel pipe shall be made with a copper-alloy fitting or dielectric fitting. The copper tubing shall be soldered to the fitting in an *approved* manner, and the fitting shall be screwed to the threaded pipe.

705.16.3 Cast-iron pipe to galvanized steel pipe. Joints between cast iron and galvanized steel shall be made by either caulked or threaded joints or with an *approved* adapter fitting.

705.16.4 Plastic pipe or tubing to other piping material. Joints between different types of plastic pipe shall be made with an *approved* adapter fitting, or by a solvent cement joint only where a single joint is made between ABS and PVC pipes at the end of a building drainage pipe and the beginning of a *building sewer* pipe using a solvent cement complying with ASTM D3138. Joints between plastic pipe and other piping material shall be made with an approved adapter fitting. Joints between plastic pipe and cast-iron hub pipe shall be made by a caulked joint or a mechanical compression joint.

705.16.5 Lead pipe to other piping material. Joints between lead pipe and other piping material shall be made by a wiped joint to a caulking ferrule, soldering nipple or bushing or shall be made with an *approved* adapter fitting.

705.16.6 Borosilicate glass to other materials. Joints between glass pipe and other types of materials shall be made with adapters having a TFE seal and shall be installed in accordance with the manufacturer's instructions.

705.16.7 Stainless steel drainage systems to other materials. Joints between stainless steel drainage systems and other piping materials shall be made with *approved* mechanical couplings.

705.17 Drainage slip joints. Slip joints shall comply with Section 405.9.

705.18 Caulking ferrules. Caulking ferrules shall be of copper alloy and shall be in accordance with Table 705.18.

TABLE 705.18
CAULKING FERRULE SPECIFICATIONS

PIPE SIZES (inches)	INSIDE DIAMETER (inches)	LENGTH (inches)	MINIMUM WEIGHT EACH
2	$2^1/_4$	$4^1/_2$	1 pound
3	$3^1/_4$	$4^1/_2$	1 pound 12 ounces
4	$4^1/_4$	$4^1/_2$	2 pounds 8 ounces

For SI: 1 inch = 25.4 mm, 1 ounce = 28.35 g, 1 pound = 0.454 kg.

705.19 Soldering bushings. Soldering bushings shall be of copper or copper alloy and shall be in accordance with Table 705.19.

TABLE 705.19
SOLDERING BUSHING SPECIFICATIONS

PIPE SIZES (inches)	MINIMUM WEIGHT EACH
$1^1/_4$	6 ounces
$1^1/_2$	8 ounces
2	14 ounces
$2^1/_2$	1 pound 6 ounces
3	2 pounds
4	3 pounds 8 ounces

For SI: 1 inch = 25.4 mm, 1 ounce = 28.35 g, 1 pound = 0.454 kg.

705.20 Stainless steel drainage systems. O-ring joints for stainless steel drainage systems shall be made with an *approved* elastomeric seal.

SECTION 706
CONNECTIONS BETWEEN DRAINAGE PIPING AND FITTINGS

706.1 Connections and changes in direction. Connections and changes in direction of the sanitary drainage system shall be made with *approved* drainage fittings. Connections between drainage piping and fixtures shall conform to Section 405.

706.2 Obstructions. The fittings shall not have ledges, shoulders or reductions capable of retarding or obstructing flow in the piping. Threaded drainage pipe fittings shall be of the recessed drainage type. This section shall not be applicable to tubular waste fittings used to convey vertical flow upstream of the trap seal liquid level of a fixture trap.

706.3 Installation of fittings. Fittings shall be installed to guide sewage and waste in the direction of flow. Change in direction shall be made by fittings installed in accordance with Table 706.3. Change in direction by combination fittings, side inlets or increasers shall be installed in accordance with Table 706.3 based on the pattern of flow created by the fitting. Double sanitary tee patterns shall not receive the discharge of back-to-back water closets and fixtures or appliances with pumping action discharge.

> **Exception:** Back-to-back water closet connections to double sanitary tees shall be permitted where the horizontal *developed length* between the outlet of the water closet and the connection to the double sanitary tee pattern is 18 inches (457 mm) or greater.

706.4 Heel- or side-inlet quarter bends. Heel-inlet quarter bends shall be an acceptable means of connection, except where the quarter bend serves a water closet. A low-heel inlet shall not be used as a wet-vented connection. Side-inlet quarter bends shall be an acceptable means of connection for drainage, wet venting and *stack* venting arrangements.

TABLE 706.3
FITTINGS FOR CHANGE IN DIRECTION

TYPE OF FITTING PATTERN	CHANGE IN DIRECTION		
	Horizontal to vertical	Vertical to horizontal	Horizontal to horizontal
Sixteenth bend	X	X	X
Eighth bend	X	X	X
Sixth bend	X	X	X
Quarter bend	X	X^a	X^a
Short sweep	X	X^{a,b}	X^a
Long sweep	X	X	X
Sanitary tee	X^c	—	—
Wye	X	X	X
Combination wye and eighth bend	X	X	X

For SI: 1 inch = 25.4 mm.

a. The fittings shall only be permitted for a 2-inch or smaller fixture drain.
b. Three inches or larger.
c. For a limitation on double sanitary tees, see Section 706.3.

SECTION 707
PROHIBITED JOINTS AND CONNECTIONS

707.1 Prohibited joints. The following types of joints and connections shall be prohibited:

1. Cement or concrete joints.
2. Mastic or hot-pour bituminous joints.
3. Joints made with fittings not *approved* for the specific installation.
4. Joints between different diameter pipes made with elastomeric rolling O-rings.
5. Solvent-cement joints between different types of plastic pipe except where provided for in Section 705.16.4.
6. Saddle-type fittings.

SECTION 708
CLEANOUTS

708.1 Cleanouts required. Cleanouts shall be provided for drainage piping in accordance with Sections 708.1.1 through 708.1.11.

708.1.1 Horizontal drains and building drains. Horizontal drainage pipes in buildings shall have cleanouts located at intervals of not more than 100 feet (30 480 mm). *Building drains* shall have cleanouts located at intervals of not more than 100 feet (30 480 mm) except where manholes are used instead of cleanouts, the manholes shall be located at intervals of not more than 400 feet (122 m). The interval length shall be measured from the cleanout or manhole opening, along the *developed length* of the piping to the next drainage fitting providing access for cleaning, the end of the horizontal drain or the end of the *building drain*.

Exception: Horizontal *fixture drain* piping serving a nonremovable trap shall not be required to have a cleanout for the section of piping between the trap and the vent connection for such trap.

708.1.2 Building sewers. *Building sewers* smaller than 8 inches (203 mm) shall have cleanouts located at intervals of not more than 100 feet (30 480 mm). *Building sewers* 8 inches (203 mm) and larger shall have a manhole located not more than 200 feet (60 960 mm) from the junction of the *building drain* and *building sewer* and at intervals of not more than 400 feet (122 m). The interval length shall be measured from the cleanout or manhole opening, along the *developed length* of the piping to the next drainage fitting providing access for cleaning, a manhole or the end of the *building sewer*.

708.1.3 Building drain and building sewer junction. The junction of the *building drain* and the *building sewer* shall be served by a cleanout that is located at the junction or within 10 feet (3048 mm) of the *developed length* of piping upstream of the junction. For the requirements of this section, the removal of the water closet shall not be required to provide cleanout access.

708.1.4 Changes of direction. Where a horizontal drainage pipe, a *building drain* or a *building sewer* has a change of horizontal direction greater than 45 degrees (0.79 rad), a cleanout shall be installed at the change of direction. Where more than one change of horizontal direction greater than 45 degrees (0.79 rad) occurs within 40 feet (12 192 mm) of *developed length* of piping, the cleanout installed for the first change of direction shall serve as the cleanout for all changes in direction within that 40 feet (12 192 mm) of *developed length* of piping.

708.1.5 Cleanout size. Cleanouts shall be the same size as the piping served by the cleanout, except that cleanouts for piping larger than 4 inches (102 mm) need not be larger than 4 inches (102 mm).

Exceptions:

1. A removable P-trap with slip or ground joint connections can serve as a cleanout for drain piping that is one size larger than the P-trap size.
2. Cleanouts located on *stacks* can be one size smaller than the *stack* size.
3. The size of cleanouts for cast-iron piping can be in accordance with the referenced standards for cast-iron fittings as indicated in Table 702.4.

708.1.6 Cleanout plugs. Cleanout plugs shall be of copper-alloy, plastic or other *approved* materials. Cleanout plugs for borosilicate glass piping systems shall be of borosilicate glass. Copper-alloy cleanout plugs shall conform to ASTM A74 and shall be limited for use only on metallic piping systems. Plastic cleanout plugs shall conform to the referenced standards for plastic pipe fittings, as indicated in Table 702.4. Cleanout plugs shall have a raised square head, a countersunk square head or a countersunk slot head. Where a cleanout plug will have a trim cover screw installed into the plug, the plug shall be manufactured with a blind end threaded hole for such purpose.

708.1.7 Manholes. Manholes and manhole covers shall be of an *approved* type. Manholes located inside of a building shall have gas-tight covers that require tools for removal.

708.1.8 Installation arrangement. The installation arrangement of a cleanout shall enable cleaning of drainage piping only in the direction of drainage flow.

Exceptions:

1. Test tees serving as cleanouts.
2. A two-way cleanout installation that is *approved* for meeting the requirements of Section 708.1.3.

708.1.9 Required clearance. Cleanouts for 6-inch (153 mm) and smaller piping shall be provided with a clearance of not less than 18 inches (457 mm) from, and perpendicular to, the face of the opening to any obstruction. Cleanouts for 8-inch (203 mm) and larger piping shall be provided with a clearance of not less than 36 inches (914 mm) from, and perpendicular to, the face of the opening to any obstruction.

708.1.10 Cleanout access. Required cleanouts shall not be installed in concealed locations. For the purposes of this section, concealed locations include, but are not limited to, the inside of plenums, within walls, within floor/ceiling assemblies, below grade and in crawl spaces where the height from the crawl space floor to the nearest obstruction along the path from the crawl space opening to the cleanout location is less than 24 inches (610 mm). Cleanouts with openings at a finished wall shall have the face of the opening located within $1^1/_2$ inches (38 mm) of the finished wall surface. Cleanouts located below grade shall be extended to grade level so that the top of the cleanout plug is at or above grade. A cleanout installed in a floor or walkway that will not have a trim cover installed shall have a countersunk plug installed so the top surface of the plug is flush with the finished surface of the floor or walkway.

708.1.10.1 Cleanout plug trim covers. Trim covers and access doors for cleanout plugs shall be designed for such purposes and shall be *approved*. Trim cover fasteners that thread into cleanout plugs shall be corrosion resistant. Cleanout plugs shall not be covered with mortar, plaster or any other permanent material.

708.1.10.2 Floor cleanout assemblies. Where it is necessary to protect a cleanout plug from the loads of vehicular traffic, cleanout assemblies in accordance with ASME A112.36.2M shall be installed.

708.1.11 Prohibited use. The use of a threaded cleanout opening to add a fixture or to extend piping shall be prohibited except where another cleanout of equal size is installed with the required access and clearance.

SECTION 709
FIXTURE UNITS

709.1 Values for fixtures. *Drainage fixture unit* values as given in Table 709.1 designate the relative load weight of different kinds of fixtures that shall be employed in estimating the total load carried by a soil or waste pipe, and shall be used in connection with Tables 710.1(1) and 710.1(2) of sizes for soil, waste and vent pipes for which the permissible load is given in terms of fixture units.

709.2 Fixtures not listed in Table 709.1. Fixtures not listed in Table 709.1 shall have a *drainage fixture unit* load based on the outlet size of the fixture in accordance with Table 709.2. The minimum trap size for unlisted fixtures shall be the size of the drainage outlet but not less than $1^1/_4$ inches (32 mm).

TABLE 709.2
DRAINAGE FIXTURE UNITS FOR FIXTURE DRAINS OR TRAPS

FIXTURE DRAIN OR TRAP SIZE (inches)	DRAINAGE FIXTURE UNIT VALUE
$1^1/_4$	1
$1^1/_2$	2
2	3
$2^1/_2$	4
3	5
4	6

For SI: 1 inch = 25.4 mm.

709.3 Conversion of gpm flow to dfu values. Where discharges to a waste receptor or to a drainage system are only known in gallons per minute (liters per second) values, the *drainage fixture unit* values for those flows shall be computed on the basis that 1 gpm (0.06 L/s) of flow is equivalent to two *drainage fixture units*.

709.4 Values for indirect waste receptor. The *drainage fixture unit* load of an indirect waste receptor receiving the discharge of indirectly connected fixtures shall be the sum of the *drainage fixture unit* values of the fixtures that discharge to the receptor, but not less than the *drainage fixture unit* value given for the indirect waste receptor in Table 709.1 or 709.2.

709.4.1 Clear-water waste receptors. Where waste receptors such as floor drains, floor sinks and hub drains receive only clear-water waste from display cases, refrigerated display cases, ice bins, coolers and freezers, such receptors shall have a *drainage fixture unit* value of one-half.

SECTION 710
DRAINAGE SYSTEM SIZING

710.1 Maximum fixture unit load. The maximum number of *drainage fixture units* connected to a given size of *building sewer*, *building drain* or horizontal *branch* of the *building drain* shall be determined using Table 710.1(1). The maximum number of drainage fixture units connected to a given size of horizontal *branch* or vertical soil or waste *stack* shall be determined using Table 710.1(2).

SANITARY DRAINAGE

TABLE 709.1
DRAINAGE FIXTURE UNITS FOR FIXTURES AND GROUPS

FIXTURE TYPE	DRAINAGE FIXTURE UNIT VALUE AS LOAD FACTORS	MINIMUM SIZE OF TRAP (inches)
Automatic clothes washers, commercial[a,g]	3	2
Automatic clothes washers, residential[g]	2	2
Bathroom group as defined in Section 202 (1.6 gpf water closet)[f]	5	—
Bathroom group as defined in Section 202 (water closet flushing greater than 1.6 gpf)[f]	6	—
Bathtub[b] (with or without overhead shower or whirlpool attachments)	2	$1\frac{1}{2}$
Bidet	1	$1\frac{1}{4}$
Combination sink and tray	2	$1\frac{1}{2}$
Dental lavatory	1	$1\frac{1}{4}$
Dental unit or cuspidor	1	$1\frac{1}{4}$
Dishwashing machine[c], domestic	2	$1\frac{1}{2}$
Drinking fountain	$\frac{1}{2}$	$1\frac{1}{4}$
Emergency floor drain	0	2
Floor drains[h]	2[h]	2
Floor sinks	Note h	2
Kitchen sink, domestic	2	$1\frac{1}{2}$
Kitchen sink, domestic with food waste disposer, dishwasher or both	2	$1\frac{1}{2}$
Laundry tray (1 or 2 compartments)	2	$1\frac{1}{2}$
Lavatory	1	$1\frac{1}{4}$
Shower (based on the total flow rate through showerheads and body sprays) Flow rate:		
5.7 gpm or less	2	$1\frac{1}{2}$
Greater than 5.7 gpm to 12.3 gpm	3	2
Greater than 12.3 gpm to 25.8 gpm	5	3
Greater than 25.8 gpm to 55.6 gpm	6	4
Service sink	2	$1\frac{1}{2}$
Sink	2	$1\frac{1}{2}$
Urinal	4	Note d
Urinal, 1 gallon per flush or less	2[e]	Note d
Urinal, nonwater supplied	$\frac{1}{2}$	Note d
Wash sink (circular or multiple) each set of faucets	2	$1\frac{1}{2}$
Water closet, flushometer tank, public or private	4[e]	Note d
Water closet, private (1.6 gpf)	3[e]	Note d
Water closet, private (flushing greater than 1.6 gpf)	4[e]	Note d
Water closet, public (1.6 gpf)	4[e]	Note d
Water closet, public (flushing greater than 1.6 gpf)	6[e]	Note d

For SI: 1 inch = 25.4 mm, 1 gallon = 3.785 L, gpf = gallon per flushing cycle, gpm = gallon per minute.

a. For traps larger than 3 inches, use Table 709.2.
b. A showerhead over a bathtub or whirlpool bathtub attachment does not increase the drainage fixture unit value.
c. See Sections 709.2 through 709.4.1 for methods of computing unit value of fixtures not listed in this table or for rating of devices with intermittent flows.
d. Trap size shall be consistent with the fixture outlet size.
e. For the purpose of computing loads on building drains and sewers, water closets and urinals shall not be rated at a lower drainage fixture unit unless the lower values are confirmed by testing.
f. For fixtures added to a bathroom group, add the dfu value of those additional fixtures to the bathroom group fixture count.
g. See Section 406.2 for sizing requirements for fixture drain, branch drain and drainage stack for an automatic clothes washer standpipe.
h. See Sections 709.4 and 709.4.1.

SANITARY DRAINAGE

TABLE 710.1(1)
BUILDING DRAINS AND SEWERS

DIAMETER OF PIPE (inches)	MAXIMUM NUMBER OF DRAINAGE FIXTURE UNITS CONNECTED TO ANY PORTION OF THE BUILDING DRAIN OR THE BUILDING SEWER, INCLUDING BRANCHES OF THE BUILDING DRAIN[a]			
	Slope per foot			
	$^1/_{16}$ inch	$^1/_8$ inch	$^1/_4$ inch	$^1/_2$ inch
$1^1/_4$	—	—	1	1
$1^1/_2$	—	—	3	3
2	—	—	21	26
$2^1/_2$	—	—	24	31
3	—	36	42	50
4	—	180	216	250
5	—	390	480	575
6	—	700	840	1,000
8	1,400	1,600	1,920	2,300
10	2,500	2,900	3,500	4,200
12	3,900	4,600	5,600	6,700
15	7,000	8,300	10,000	12,000

For SI: 1 inch = 25.4 mm, 1 inch per foot = 83.3 mm/m.

a. The minimum size of any building drain serving a water closet shall be 3 inches.

TABLE 710.1(2)
HORIZONTAL FIXTURE BRANCHES AND STACKS[a]

DIAMETER OF PIPE (inches)	MAXIMUM NUMBER OF DRAINAGE FIXTURE UNITS (dfu)			
	Total for horizontal branch	Stacks[b]		
		Total discharge into one branch interval	Total for stack of three branch Intervals or less	Total for stack greater than three branch intervals
$1^1/_2$	3	2	4	8
2	6	6	10	24
$2^1/_2$	12	9	20	42
3	20	20	48	72
4	160	90	240	500
5	360	200	540	1,100
6	620	350	960	1,900
8	1,400	600	2,200	3,600
10	2,500	1,000	3,800	5,600
12	3,900	1,500	6,000	8,400
15	7,000	Note c	Note c	Note c

For SI: 1 inch = 25.4 mm.

a. Does not include branches of the building drain. Refer to Table 710.1(1).
b. Stacks shall be sized based on the total accumulated connected load at each story or branch interval. As the total accumulated connected load decreases, stacks are permitted to be reduced in size. Stack diameters shall not be reduced to less than one-half of the diameter of the largest stack size required.
c. Sizing load based on design criteria.

710.1.1 Horizontal stack offsets. Horizontal *stack* offsets shall be sized as required for building drains in accordance with Table 710.1(1), except as required by Section 711.3.

710.1.2 Vertical stack offsets. Vertical *stack* offsets shall be sized as required for straight *stacks* in accordance with Table 710.1(2), except where required to be sized as a *building drain* in accordance with Section 711.1.1.

710.2 Future fixtures. Where provision is made for the future installation of fixtures, those provided for shall be considered in determining the required sizes of drain pipes.

SECTION 711
OFFSETS IN DRAINAGE PIPING IN BUILDINGS OF FIVE STORIES OR MORE

711.1 Horizontal branch connections above or below vertical stack offsets. If a horizontal *branch* connects to the *stack* within 2 feet (610 mm) above or below a vertical *stack* offset, and the offset is located more than four *branch intervals* below the top of the *stack*, the offset shall be vented in accordance with Section 907.

711.1.1 Omission of vents for vertical stack offsets. Vents for vertical offsets required by Section 711.1 shall not be required where the *stack* and its offset are sized as a *building drain* [see Table 710.1(1)].

711.2 Horizontal stack offsets. A *stack* with a horizontal offset located more than four *branch intervals* below the top of the *stack* shall be vented in accordance with Section 907 and sized as follows:

1. The portion of the *stack* above the offset shall be sized as for a vertical *stack* based on the total number of *drainage fixture units* above the offset.

2. The offset shall be sized in accordance with Section 710.1.1.

3. The portion of the *stack* below the offset shall be sized as for the offset or based on the total number of *drainage fixture units* on the entire *stack*, whichever is larger [see Table 710.1(2), Column 5].

711.2.1 Omission of vents for horizontal stack offsets. Vents for horizontal *stack* offsets required by Section 711.2 shall not be required where the *stack* and its offset are one pipe size larger than required for a building drain [see Table 710.1(1)] and the entire *stack* and offset are not less in cross-sectional area than that required for a straight stack plus the area of an offset vent as provided for in Section 907.

711.3 Offsets below lowest branch. Where a vertical offset occurs in a soil or waste *stack* below the lowest horizontal *branch*, a change in diameter of the *stack* because of the offset shall not be required. If a horizontal offset occurs in a soil or waste *stack* below the lowest horizontal *branch*, the required diameter of the offset and the *stack* below it shall be determined as for a *building drain* in accordance with Table 710.1(1).

SECTION 712
SUMPS AND EJECTORS

712.1 Building subdrains. *Building subdrains* that cannot be discharged to the *sewer* by gravity flow shall be discharged into a tightly covered and vented sump from which the liquid shall be lifted and discharged into the building gravity drainage system by automatic pumping equipment or other *approved* method. In other than existing structures, the sump shall not receive drainage from any piping within the building capable of being discharged by gravity to the *building sewer*.

712.2 Valves required. A check valve and a full open valve located on the discharge side of the check valve shall be installed in the pump or ejector discharge piping between the pump or ejector and the gravity drainage system. *Access* shall be provided to such valves. Such valves shall be located above the sump cover required by Section 712.1 or, where the discharge pipe from the ejector is below grade, the valves shall be accessibly located outside the sump below grade in an access pit with a removable *access* cover.

712.3 Sump design. The sump pump, pit and discharge piping shall conform to the requirements of Sections 712.3.1 through 712.3.5.

712.3.1 Sump pump. The sump pump capacity and head shall be appropriate to anticipated use requirements.

712.3.2 Sump pit. The sump pit shall be not less than 18 inches (457 mm) in diameter and not less than 24 inches (610 mm) in depth, unless otherwise *approved*. The pit shall be provided with *access* and shall be located such that all drainage flows into the pit by gravity. The sump pit shall be constructed of tile, concrete, steel, plastic or other approved materials. The pit bottom shall be solid and provide permanent support for the pump. The sump pit shall be fitted with a gastight removable cover that is installed not more than 2 inches (51 mm) below grade or floor level. The cover shall be adequate to support anticipated loads in the area of use. The sump pit shall be vented in accordance with Chapter 9.

712.3.3 Discharge pipe and fittings. Discharge pipe and fittings serving sump pumps and ejectors shall be constructed of materials in accordance with Sections 712.3.3.1 and 712.3.3.2.

712.3.3.1 Materials. Pipe and fitting materials shall be constructed of copper or copper-alloy, CPVC, ductile iron, PE, or PVC.

712.3.3.2 Ratings. Pipe and fittings shall be rated for the maximum system operating pressure and temperature. Pipe fitting materials shall be compatible with the pipe material. Where pipe and fittings are buried in the earth, they shall be suitable for burial.

712.3.4 Maximum effluent level. The effluent level control shall be adjusted and maintained to at all times prevent the effluent in the sump from rising to within 2 inches (51 mm) of the invert of the gravity drain inlet into the sump.

712.3.5 Pump connection to the drainage system. Pumps connected to the drainage system shall connect to a *building sewer*, building drain, soil *stack*, waste *stack* or *horizontal branch drain*. Where the discharge line connects into horizontal drainage piping, the connection shall be made through a wye fitting into the top of the drainage piping and such wye fitting shall be located not less than 10 pipe diameters from the base of any soil *stack*, waste *stack* or *fixture drain*.

712.4 Sewage pumps and sewage ejectors. A sewage pump or sewage ejector shall automatically discharge the contents of the sump to the building drainage system.

712.4.1 Macerating toilet systems. Macerating toilet systems shall comply with ASME A112.3.4/CSA B45.9 and shall be installed in accordance with the manufacturer's instructions.

712.4.2 Capacity. A sewage pump or sewage ejector shall have the capacity and head for the application requirements. Pumps or ejectors that receive the discharge of water closets shall be capable of handling spherical solids with a diameter of up to and including 2 inches (51 mm). Other pumps or ejectors shall be capable of handling spherical solids with a diameter of up to and including $^1/_2$ inch (13 mm). The capacity of a pump or ejector based on the diameter of the discharge pipe shall be not less than that indicated in Table 712.4.2.

Exceptions:

1. Grinder pumps or grinder ejectors that receive the discharge of water closets shall have a discharge opening of not less than $1^1/_4$ inches (32 mm).

2. Macerating toilet assemblies that serve single water closets shall have a discharge opening of not less than $^3/_4$ inch (19.1 mm).

TABLE 712.4.2
MINIMUM CAPACITY OF SEWAGE PUMP OR SEWAGE EJECTOR

DIAMETER OF THE DISCHARGE PIPE (inches)	CAPACITY OF PUMP OR EJECTOR (gpm)
2	21
$2^1/_2$	30
3	46

For SI: 1 inch = 25.4 mm, 1 gallon per minute = 3.785 L/m.

SECTION 713
COMPUTERIZED DRAINAGE DESIGN

713.1 Design of drainage system. The sizing, design and layout of the drainage system shall be permitted to be designed by *approved* computer design methods.

713.2 Load on drainage system. The load shall be computed from the simultaneous or sequential discharge conditions from fixtures, appurtenances and appliances or the peak usage design condition.

713.2.1 Fixture discharge profiles. The discharge profiles for flow rates versus time from fixtures and appliances shall be in accordance with the manufacturer's specifications.

713.3 Selections of drainage pipe sizes. Pipe shall be sized to prevent full-bore flow.

713.3.1 Selecting pipe wall roughness. Pipe size calculations shall be conducted with the pipe wall roughness factor (k_s), in accordance with the manufacturer's specifications and as modified for aging roughness factors with deposits and corrosion.

713.3.2 Slope of horizontal drainage piping. Horizontal drainage piping shall be designed and installed at slopes in accordance with Table 704.1.

SANITARY DRAINAGE

SECTION 714
BACKWATER VALVES

714.1 Sewage backflow. Where plumbing fixtures are installed on a floor with a finished floor elevation below the elevation of the manhole cover of the next upstream manhole in the public *sewer,* such fixtures shall be protected by a backwater valve installed in the *building drain,* or horizontal *branch* serving such fixtures. Plumbing fixtures installed on a floor with a finished floor elevation above the elevation of the manhole cover of the next upstream manhole in the public *sewer* shall not discharge through a backwater valve.

> **Exception:** In existing buildings, fixtures above the elevation of the manhole cover of the next upstream manhole in the public *sewer* shall not be prohibited from discharging through a backwater valve.

714.2 Material. Backwater valves shall comply with ASME A112.14.1, CSA B181.1 or CSA B181.2.

714.3 Location. Backwater valves shall be installed so that *access* is provided to the working parts.

SECTION 715
VACUUM DRAINAGE SYSTEMS

715.1 Scope. Vacuum drainage systems shall be in accordance with Sections 715.2 through 715.4.

715.2 System design. Vacuum drainage systems shall be designed in accordance with the vacuum drainage system manufacturer's instructions. The system layout, including piping layout, tank assemblies, vacuum pump assembly and other components necessary for proper function of the system shall be in accordance with the manufacturer's instructions. Plans, specifications and other data for such systems shall be submitted to the code official for review and approval prior to installation.

715.2.1 Fixtures. Gravity-type fixtures installed in vacuum drainage systems shall comply with Chapter 4.

715.2.2 Drainage fixture units. *Drainage fixture units* for gravity drainage systems that discharge into, or receive discharge from, vacuum drainage systems shall be based on the values in this chapter.

715.2.3 Water supply fixture units. Water supply fixture units shall be based on the values in Chapter 6 of this code, except that the water supply fixture unit for a vacuum-type water closet shall be 1.

715.2.4 Traps and cleanouts. Gravity drainage fixtures shall be provided with traps and cleanouts in accordance with this chapter and Chapter 10.

715.2.5 Materials. Vacuum drainage pipe, fitting and valve materials shall be in accordance with the vacuum drainage system manufacturer's instructions and the requirements of this chapter.

715.3 Testing and demonstrations. After completion of the entire system installation, the system shall be subjected to a vacuum test of 19 inches (483 mm) of mercury and shall be operated to function as required by the code official and the manufacturer of the vacuum drainage system. Recorded proof of all tests shall be submitted to the code official.

715.4 Written instructions. Written instructions for the operation, maintenance, safety and emergency procedures shall be provided to the building owner. The code official shall verify that the building owner is in receipt of such instructions.

SECTION 716
REPLACEMENT OF UNDERGROUND BUILDING SEWERS AND BUILDING DRAINS BY PIPE-BURSTING METHODS

716.1 General. This section shall govern the replacement of existing *building sewer* and *building drain* piping by pipe-bursting methods.

716.2 Applicability. The replacement of *building sewer* and *building drain* piping by pipe-bursting methods shall be limited to gravity drainage piping of sizes 6 inches (152 mm) and smaller. The replacement piping shall be of the same nominal size as the existing piping.

716.3 Pre-installation inspection. The existing piping sections to be replaced shall be inspected internally by a recorded video camera survey. The survey shall include notations of the position of cleanouts and the depth of connections to the existing piping.

716.4 Pipe. The replacement pipe shall be made of high-density polyethylene (HDPE) and shall have a standard dimension ratio (SDR) of 17. The pipe shall be in compliance with ASTM F714.

716.5 Pipe fittings. Pipe fittings to be connected to the replacement pipe shall be made of high-density polyethylene (HDPE) and shall be in compliance with ASTM D2683.

716.6 Cleanouts. Where the existing *building sewer* or *building drain* did not have cleanouts meeting the requirements of this code, cleanout fittings shall be installed as required by this code.

716.7 Post-installation inspection. The completed replacement piping section shall be inspected internally by a recorded video camera survey. The video survey shall be reviewed and *approved* by the code official prior to pressure testing of the replacement piping system.

716.8 Pressure testing. The replacement piping system as well as the connections to the replacement piping shall be tested in accordance with Section 312.

CHAPTER 8

INDIRECT/SPECIAL WASTE

User note:

About this chapter: There are drainage applications in buildings where a backup of liquid waste in a drainage system could contaminate equipment and appliances. Chapter 8 covers the applications that require an indirect discharge connection to the building's drainage system. The chapter has provisions for the types of indirect connections and waste receptor configurations.

SECTION 801
GENERAL

801.1 Scope. This chapter shall govern matters concerning indirect waste piping and special wastes. This chapter shall further control matters concerning food-handling establishments, sterilizers, humidifiers, clear-water waste, swimming pools, methods of providing *air breaks* or *air gaps*, and neutralizing devices for corrosive wastes.

801.2 Protection. Devices, appurtenances, appliances and apparatus intended to serve some special function, such as sterilization, humidification, distillation, processing, cooling, or storage of ice or foods, and that discharge to the drainage system, shall be provided with protection against backflow, flooding, fouling, contamination and stoppage of the drain.

SECTION 802
INDIRECT WASTES

802.1 Where required. Food-handling equipment, in other than dwelling units, clear-water waste, humidifiers, dishwashing machines and utensils, pots, pans and dishwashing sinks shall discharge through an indirect waste pipe as specified in Sections 802.1.1 through 802.1.7. Fixtures not required to be indirectly connected by this section and the exception to Section 301.6 shall be directly connected to the plumbing system in accordance with Chapter 7.

802.1.1 Food handling. Equipment and fixtures utilized for the storage, preparation and handling of food shall discharge through an indirect waste pipe by means of an *air gap*. Each well of a multiple-compartment sink shall discharge independently to a waste receptor.

802.1.2 Floor drains in food storage areas. Floor drains located within walk-in refrigerators or freezers in food service and food establishments shall be indirectly connected to the sanitary drainage system by means of an *air gap*. Where a floor drain is located within an area subject to freezing, the waste line serving the floor drain shall not be trapped and shall indirectly discharge into a waste receptor located outside of the area subject to freezing.

> **Exception:** Where protected against backflow by a backwater valve, such floor drains shall be indirectly connected to the sanitary drainage system by means of an *air break* or an *air gap*.

802.1.3 Potable clear-water waste. Where devices and equipment, such as sterilizers and relief valves, discharge potable water to the building drainage system, the discharge shall be through an indirect waste pipe by means of an *air gap*.

802.1.4 Swimming pools. Where wastewater from swimming pools, backwash from filters and water from pool deck drains discharge to the building drainage system, the discharge shall be through an indirect waste pipe by means of an *air gap*.

802.1.5 Nonpotable clear-water waste. Where devices and equipment such as process tanks, filters, drips and boilers discharge nonpotable water to the building drainage system, the discharge shall be through an indirect waste pipe by means of an *air break* or an *air gap*.

802.1.6 Commercial dishwashing machines. The discharge from a commercial dishwashing machine shall be through an *air gap* or *air break* into a waste receptor in accordance with Section 802.3.

802.1.7 Food utensils, dishes, pots and pans sinks. Sinks, in other than dwelling units, used for the washing, rinsing or sanitizing of utensils, dishes, pots, pans or service ware used in the preparation, serving or eating of food shall discharge indirectly through an *air gap* or an *air break* to the drainage system.

802.2 Material, joints and connections. The materials, joints, connections and methods utilized for the construction and installation of indirect waste piping systems shall comply with the applicable provisions of Chapter 7.

802.3 Installation. Indirect waste piping shall discharge through an *air gap* or *air break* into a waste receptor. Waste receptors shall be trapped and vented and shall connect to the building drainage system. Indirect waste piping that exceeds 30 inches (762 mm) in *developed length* measured horizontally, or 54 inches (1372 mm) in total *developed length*, shall be trapped.

> **Exception:** Where a waste receptor receives only clearwater waste and does not directly connect to a sanitary drainage system, the receptor shall not require a trap.

802.3.1 Air gap. The *air gap* between the indirect waste pipe and the *flood level rim* of the waste receptor shall be not less than twice the effective opening of the indirect waste pipe.

802.3.2 Air break. An *air break* shall be provided between the indirect waste pipe and the trap seal of the waste receptor.

802.4 Waste receptors. For other than hub drains that receive only clear-water waste and standpipes, a removable strainer or basket shall cover the outlet of waste receptors. Waste receptors shall not be installed in concealed spaces. Waste receptors shall not be installed in plenums, crawl spaces, attics, interstitial spaces above ceilings and below floors. Ready *access* shall be provided to waste receptors.

802.4.1 Size of receptors. A waste receptor shall be sized for the maximum discharge of all indirect waste pipes served by the receptor. Receptors shall be installed to prevent splashing or flooding.

802.4.2 Hub drains. A hub drain shall be in the form of a hub or a pipe extending not less than 1 inch (25 mm) above a water-impervious floor.

802.4.3 Standpipes. Standpipes shall be individually trapped. Standpipes shall extend not less than 18 inches (457 mm) but not greater than 42 inches (1066 mm) above the trap weir. *Access* shall be provided to standpipes and drains for rodding.

802.4.3.1 Connection of laundry tray to standpipe. As an alternative for a laundry tray fixture connecting directly to a drainage system, a laundry tray waste line without a fixture trap shall connect to a standpipe for an automatic clothes washer drain. The standpipe shall extend not less than 30 inches (732 mm) above the weir of the standpipe trap and shall extend above the *flood level rim* of the laundry tray. The outlet of the laundry tray shall not be greater than 30 inches (762 mm) horizontal distance from the side of the standpipe.

SECTION 803
SPECIAL WASTES

803.1 Neutralizing device required for corrosive wastes. Corrosive liquids, spent acids or other harmful chemicals that destroy or injure a drain, *sewer*, soil or waste pipe, or create noxious or toxic fumes or interfere with sewage treatment processes shall not be discharged into the plumbing system without being thoroughly diluted, neutralized or treated by passing through an *approved* dilution or neutralizing device. Such devices shall be automatically provided with a sufficient supply of diluting water or neutralizing medium so as to make the contents noninjurious before discharge into the drainage system. The nature of the corrosive or harmful waste and the method of its treatment or dilution shall be *approved* prior to installation.

803.2 System design. A chemical drainage and vent system shall be designed and installed in accordance with this code. Chemical drainage and vent systems shall be completely separated from the sanitary systems. Chemical waste shall not discharge to a sanitary drainage system until such waste has been treated in accordance with Section 803.1.

CHAPTER 9
VENTS

User note:

About this chapter: Chapter 9 regulates connection locations, various venting system arrangements and the sizing of piping for vent systems. The proper operation of a gravity flow drainage system (Chapter 7) depends on maintaining an air path throughout the system to prevent waste and odor "blow back" into fixtures and siphoning of the trap seal in fixture traps (Chapter 10).

SECTION 901
GENERAL

901.1 Scope. The provisions of this chapter shall govern the materials, design, construction and installation of vent systems.

901.2 Trap seal protection. The plumbing system shall be provided with a system of vent piping that will permit the admission or emission of air so that the seal of any fixture trap shall not be subjected to a pressure differential of more than 1 inch of water column (249 Pa).

901.2.1 Venting required. Traps and trapped fixtures shall be vented in accordance with one of the venting methods specified in this chapter.

901.3 Chemical waste vent systems. The vent system for a chemical waste system shall be independent of the sanitary vent system and shall terminate separately through the roof to the outdoors or to an air admittance valve that complies with ASSE 1049. Air admittance valves for chemical waste systems shall be constructed of materials *approved* in accordance with Section 702.6 and shall be tested for chemical resistance in accordance with ASTM F1412.

901.4 Use limitations. The plumbing vent system shall not be utilized for purposes other than the venting of the plumbing system.

901.5 Tests. The vent system shall be tested in accordance with Section 312.

901.6 Engineered systems. Engineered venting systems shall conform to the provisions of Section 919.

SECTION 902
MATERIALS

902.1 Vents. The materials and methods utilized for the construction and installation of venting systems shall comply with the applicable provisions of Section 702.

902.2 Sheet copper. Sheet copper for vent pipe flashings shall conform to ASTM B152 and shall weigh not less than 8 ounces per square foot (2.5 kg/m^2).

902.3 Sheet lead. Sheet lead for vent pipe flashings shall weigh not less than 3 pounds per square foot (15 kg/m^2) for field-constructed flashings and not less than 2$^1/_2$ pounds per square foot (12 kg/m^2) for prefabricated flashings.

SECTION 903
VENT TERMINALS

903.1 Roof extension. Open vent pipes that extend through a roof shall be terminated not less than [NUMBER] inches (mm) above the roof. Where a roof is to be used for assembly or as a promenade, observation deck, sunbathing deck or similar purposes, open vent pipes shall terminate not less than 7 feet (2134 mm) above the roof.

903.2 Frost closure. Where the 97.5-percent value for outdoor design temperature is 0°F (-18°C) or less, vent extensions through a roof or wall shall be not less than 3 inches (76 mm) in diameter. Any increase in the size of the vent shall be made not less than 1 foot (305 mm) inside the thermal envelope of the building.

903.3 Flashings. The juncture of each vent pipe with the roof line shall be made watertight by an *approved* flashing.

903.4 Prohibited use. A vent terminal shall not be used for any purpose other than a vent terminal.

903.5 Location of vent terminal. An open vent terminal from a drainage system shall not be located directly beneath any door, openable window, or other air intake opening of the building or of an adjacent building, and any such vent terminal shall not be within 10 feet (3048 mm) horizontally of such an opening unless it is 3 feet (914 mm) or more above the top of such opening.

903.6 Extension through the wall. Vent terminals extending through the wall shall terminate at a point not less than 10 feet (3048 mm) from a lot line and not less than 10 feet (3048 mm) above average ground level. Vent terminals shall not terminate under the overhang of a structure with soffit vents. Side wall vent terminals shall be protected to prevent birds or rodents from entering or blocking the vent opening.

903.7 Extension outside a structure. In climates where the 97.5-percent value for outside design temperature is less than 0°F (-18°C), vent pipes installed on the exterior of the structure shall be protected against freezing by insulation, heat or both.

SECTION 904
OUTDOOR VENT EXTENSIONS

904.1 Required vent extension. The vent system serving each *building drain* shall have not less than one vent pipe that extends to the outdoors.

904.1.1 Installation. The required vent shall be a dry vent that connects to the *building drain* or an extension of a drain that connects to the *building drain*. Such vent shall not be an island fixture vent as allowed by Section 916.

904.1.2 Size. The required vent shall be sized in accordance with Section 906.2 based on the required size of the *building drain*.

904.2 Vent stack required. A vent *stack* shall be required for every drainage *stack* that has five *branch intervals* or more.

> **Exception:** Drainage stacks installed in accordance with Section 913.

904.3 Vent termination. Vent *stacks* or *stack vents* shall terminate outdoors to the open air or to a stack-type air admittance valve in accordance with Section 918.

904.4 Vent connection at base. Vent *stacks* shall connect to the base of the drainage *stack*. The vent *stack* shall connect at or below the lowest horizontal *branch*. Where the vent *stack* connects to the *building drain*, the connection shall be located downstream of the drainage *stack* and within a distance of 10 times the diameter of the drainage *stack*.

904.5 Vent headers. *Stack vents* and vent stacks connected into a common vent header at the top of the *stacks* and extending to the open air at one point shall be sized in accordance with the requirements of Section 906.1. The number of fixture units shall be the sum of all fixture units on all *stacks* connected thereto, and the *developed length* shall be the longest vent length from the intersection at the base of the most distant *stack* to the vent terminal in the open air, as a direct extension of one *stack*.

SECTION 905
VENT CONNECTIONS AND GRADES

905.1 Connection. Individual, *branch* and circuit vents shall connect to a vent *stack*, *stack vent*, air admittance valve or extend to the open air.

905.2 Grade. Vent and *branch* vent pipes shall be so graded and connected as to drain back to the drainage pipe by gravity.

905.3 Vent connection to drainage system. Every dry vent connecting to a horizontal drain shall connect above the centerline of the horizontal drain pipe.

905.4 Vertical rise of vent. Every dry vent shall rise vertically to a point not less than 6 inches (152 mm) above the *flood level rim* of the highest trap or trapped fixture being vented.

> **Exception:** Vents for interceptors located outdoors.

905.5 Height above fixtures. A connection between a vent pipe and a vent *stack* or *stack vent* shall be made at not less than 6 inches (152 mm) above the *flood level rim* of the highest fixture served by the vent. Horizontal vent pipes forming *branch* vents, relief vents or loop vents shall be located not less than 6 inches (152 mm) above the *flood level rim* of the highest fixture served.

905.6 Vent for future fixtures. Where the drainage piping has been roughed-in for future fixtures, a rough-in connection for a vent shall be installed. The vent size shall be not less than one-half the diameter of the rough-in drain to be served. The vent rough-in shall connect to the vent system, or shall be vented by other means as provided for in this chapter. The connection shall be identified to indicate that it is a vent.

SECTION 906
VENT PIPE SIZING

906.1 Size of stack vents and vent stacks. The minimum required diameter of *stack vents* and vent *stacks* shall be determined from the *developed length* and the total of *drainage fixture units* connected thereto in accordance with Table 906.1, but in no case shall the diameter be less than one-half the diameter of the drain served or less than $1^1/_4$ inches (32 mm).

906.2 Vents other than stack vents or vent stacks. The diameter of individual vents, *branch* vents, circuit vents and relief vents shall be not less than one-half the required diameter of the drain served. The required size of the drain shall be determined in accordance with Table 710.1(2). Vent pipes shall be not less than $1^1/_4$ inches (32 mm) in diameter. Vents exceeding 40 feet (12 192 mm) in *developed length* shall be increased by one nominal pipe size for the entire *developed length* of the vent pipe. Relief vents for soil and waste *stacks* in buildings having more than 10 *branch intervals* shall be sized in accordance with Section 908.2.

906.3 Developed length. The *developed length* of individual, *branch*, circuit and relief vents shall be measured from the farthest point of vent connection to the drainage system to the point of connection to the vent *stack*, *stack vent* or termination outside of the building.

906.4 Multiple branch vents. Where multiple *branch* vents are connected to a common *branch* vent, the common *branch* vent shall be sized in accordance with this section based on the size of the common horizontal drainage *branch* that is or would be required to serve the total *drainage fixture unit* load being vented.

906.5 Sump vents. Sump vent sizes shall be determined in accordance with Sections 906.5.1 and 906.5.2.

> **906.5.1 Sewage pumps and sewage ejectors other than pneumatic.** Drainage piping below *sewer* level shall be vented in the same manner as that of a gravity system. Building sump vent sizes for sumps with sewage pumps or sewage ejectors, other than pneumatic, shall be determined in accordance with Table 906.5.1.

> **906.5.2 Pneumatic sewage ejectors.** The air pressure relief pipe from a pneumatic sewage ejector shall be connected to an independent vent *stack* terminating as required for vent extensions through the roof. The relief pipe shall be sized to relieve air pressure inside the ejector to atmospheric pressure, but shall be not less than $1^1/_4$ inches (32 mm) in size.

TABLE 906.1
SIZE AND DEVELOPED LENGTH OF STACK VENTS AND VENT STACKS

DIAMETER OF SOIL OR WASTE STACK (inches)	TOTAL FIXTURE UNITS BEING VENTED (dfu)	MAXIMUM DEVELOPED LENGTH OF VENT (feet)[a] DIAMETER OF VENT (inches)										
		1¼	1½	2	2½	3	4	5	6	8	10	12
1¼	2	30	—									
1½	8	50	150	—	—	—	—	—	—	—	—	—
1½	10	30	100									
2	12	—	75	200	—							
2	20	30	50	150	—	—	—	—	—	—	—	—
2½	42	26	30	100	300							
3	10	•	42	150	360	1,040						
3	21	—	32	110	270	810	—	—	—	—	—	—
3	53		27	94	230	680						
3	102		25	86	210	620	—					
4	43	—	—	35	85	250	980	—				
4	140		—	27	65	200	750					
4	320			23	55	170	640	—				
4	540	—	—	21	50	150	580	—				
5	190			—	28	82	320	990				
5	490				21	63	250	760				
5	940	—	—	—	18	53	210	670	—	—	—	—
5	1,400				16	49	190	590				
6	500					33	130	400	1,000			
6	1,100	—	—	—	—	26	100	310	780	—	—	—
6	2,000					22	84	260	660			
6	2,900					20	77	240	600	—		
8	1,800	—	—	—	—	—	31	95	240	940	—	—
8	3,400					—	24	73	190	729		
8	5,600						20	62	160	610	—	
8	7,600	—	—	—	—	—	18	56	140	560	—	—
10	4,000						—	31	78	310	960	
10	7,200							24	60	240	740	
10	11,000	—	—	—	—	—	—	20	51	200	630	—
10	15,000							18	46	180	571	
12	7,300								31	120	380	940
12	13,000	—	—	—	—	—	—	—	24	94	300	720
12	20,000								20	79	250	610
12	26,000								18	72	230	500
15	15,000	—	—	—	—	—	—	—	—	40	130	310
15	25,000									31	96	240
15	38,000	—	—	—	—	—	—	—		26	81	200
15	50,000									24	74	180

For SI: 1 inch = 25.4 mm, 1 foot = 304.8 mm.

a. The developed length shall be measured from the vent connection to the open air.

TABLE 906.5.1
SIZE AND LENGTH OF SUMP VENTS

DISCHARGE CAPACITY OF PUMP (gpm)	MAXIMUM DEVELOPED LENGTH OF VENT (feet)[a]					
	Diameter of vent (inches)					
	1¼	1½	2	2½	3	4
10	No limit[b]	No limit	No limit	No limit	No limit	No limit
20	270	No limit	No limit	No limit	No limit	No limit
40	72	160	No limit	No limit	No limit	No limit
60	31	75	270	No limit	No limit	No limit
80	16	41	150	380	No limit	No limit
100	10[c]	25	97	250	No limit	No limit
150	Not permitted	10[c]	44	110	370	No limit
200	Not permitted	Not permitted	20	60	210	No limit
250	Not permitted	Not permitted	10	36	132	No limit
300	Not permitted	Not permitted	10[c]	22	88	380
400	Not permitted	Not permitted	Not permitted	10[c]	44	210
500	Not permitted	Not permitted	Not permitted	Not permitted	24	130

For SI: 1 inch = 25.4 mm, 1 foot = 304.8 mm, 1 gallon per minute = 3.785 L/m.

a. Developed length plus an appropriate allowance for entrance losses and friction due to fittings, changes in direction and diameter. Suggested allowances shall be obtained from NBS Monograph 31 or other approved sources. An allowance of 50 percent of the developed length shall be assumed if a more precise value is not available.
b. Actual values greater than 500 feet.
c. Less than 10 feet.

SECTION 907
VENTS FOR STACK OFFSETS

907.1 Vent for horizontal offset of drainage stack. Horizontal offsets of drainage *stacks* shall be vented where five or more *branch intervals* are located above the offset. The offset shall be vented by venting the upper section of the drainage *stack* and the lower section of the drainage *stack*.

907.2 Upper section. The upper section of the drainage *stack* shall be vented as a separate *stack* with a vent *stack* connection installed in accordance with Section 904.4. The offset shall be considered to be the base of the *stack*.

907.3 Lower section. The lower section of the drainage *stack* shall be vented by a yoke vent connecting between the offset and the next lower horizontal *branch*. The yoke vent connection shall be permitted to be a vertical extension of the drainage *stack*. The size of the yoke vent and connection shall be not less than the size required for the vent *stack* of the drainage *stack*.

SECTION 908
RELIEF VENTS—STACKS OF MORE THAN 10 BRANCH INTERVALS

908.1 Where required. Soil and waste *stacks* in buildings having more than 10 *branch intervals* shall be provided with a relief vent at each tenth interval installed, beginning with the top floor.

908.2 Size and connection. The size of the relief vent shall be equal to the size of the vent *stack* to which it connects. The lower end of each relief vent shall connect to the soil or waste *stack* through a wye below the horizontal *branch* serving the floor, and the upper end shall connect to the vent *stack* through a wye not less than 3 feet (914 mm) above the floor.

SECTION 909
FIXTURE VENTS

909.1 Distance of trap from vent. Each fixture trap shall have a protecting vent located so that the slope and the *developed length* in the *fixture drain* from the trap weir to the vent fitting are within the requirements set forth in Table 909.1.

Exception: The *developed length* of the *fixture drain* from the trap weir to the vent fitting for self-siphoning fixtures, such as water closets, shall not be limited.

TABLE 909.1
MAXIMUM DISTANCE OF FIXTURE TRAP FROM VENT

SIZE OF TRAP (inches)	SLOPE (inch per foot)	DISTANCE FROM TRAP (feet)
1¼	¼	5
1½	¼	6
2	¼	8
3	⅛	12
4	⅛	16

For SI: 1 inch = 25.4 mm, 1 foot = 304.8 mm, 1 inch per foot = 83.3 mm/m.

909.2 Venting of fixture drains. The total fall in a *fixture drain* due to pipe slope shall not exceed the diameter of the *fixture drain*, nor shall the vent connection to a *fixture drain*, except for water closets, be below the weir of the trap.

909.3 Crown vent. A vent shall not be installed within two pipe diameters of the trap weir.

SECTION 910
INDIVIDUAL VENT

910.1 Individual vent permitted. Each trap and trapped fixture is permitted to be provided with an individual vent. The individual vent shall connect to the *fixture drain* of the trap or trapped fixture being vented.

SECTION 911
COMMON VENT

911.1 Individual vent as common vent. An individual vent is permitted to vent two traps or trapped fixtures as a common vent. The traps or trapped fixtures being common vented shall be located on the same floor level.

911.2 Connection at the same level. Where the *fixture drains* being common vented connect at the same level, the vent connection shall be at the interconnection of the *fixture drains* or downstream of the interconnection.

911.3 Connection at different levels. Where the *fixture drains* connect at different levels, the vent shall connect as a vertical extension of the vertical drain. The vertical drain pipe connecting the two *fixture drains* shall be considered to be the vent for the lower *fixture drain*, and shall be sized in accordance with Table 911.3. The upper fixture shall not be a water closet.

TABLE 911.3
COMMON VENT SIZES

PIPE SIZE (inches)	MAXIMUM DISCHARGE FROM UPPER FIXTURE DRAIN (dfu)
1 1/2	1
2	4
2 1/2 to 3	6

For SI: 1 inch = 25.4 mm.

SECTION 912
WET VENTING

912.1 Horizontal wet vent permitted. Any combination of fixtures within two *bathroom groups* located on the same floor level is permitted to be vented by a horizontal wet vent. The wet vent shall be considered to be the vent for the fixtures and shall extend from the connection of the dry vent along the direction of the flow in the drain pipe to the most downstream *fixture drain* connection to the *horizontal branch drain*. Each wet-vented *fixture drain* shall connect independently to the horizontal wet vent. Only the fixtures within the *bathroom groups* shall connect to the wet-vented *horizontal branch drain*. Any additional fixtures shall discharge downstream of the horizontal wet vent.

912.1.1 Vertical wet vent permitted. Any combination of fixtures within two *bathroom groups* located on the same floor level is permitted to be vented by a vertical wet vent. The vertical wet vent shall be considered to be the vent for the fixtures and shall extend from the connection of the dry vent down to the lowest *fixture drain* connection. Each wet-vented fixture shall connect independently to the vertical wet vent. Water closet drains shall connect at the same elevation. Other *fixture drains* shall connect above or at the same elevation as the water closet *fixture drains*.

The dry-vent connection to the vertical wet vent shall be an individual or common vent serving one or two fixtures.

912.2 Dry vent connection. The required dry-vent connection for wet-vented systems shall comply with Sections 912.2.1 and 912.2.2.

912.2.1 Horizontal wet vent. The dry-vent connection for a horizontal wet-vent system shall be an individual vent or a common vent for any *bathroom group* fixture, except an *emergency floor drain*. Where the dry-vent connects to a water closet *fixture drain*, the drain shall connect horizontally to the horizontal wet-vent system. Not more than one wet-vented *fixture drain* shall discharge upstream of the dry-vented *fixture drain* connection.

912.2.2 Vertical wet vent. The dry-vent connection for a vertical wet-vent system shall be an individual vent or common vent for the most upstream *fixture drain*.

912.3 Size. The dry vent serving the wet vent shall be sized based on the largest required diameter of pipe within the wet-vent system served by the dry vent. The wet vent shall be of a size not less than that specified in Table 912.3, based on the fixture unit discharge to the wet vent.

TABLE 912.3
WET VENT SIZE

WET VENT PIPE SIZE (inches)	DRAINAGE FIXTURE UNIT LOAD (dfu)
1 1/2	1
2	4
2 1/2	6
3	12

For SI: 1 inch = 25.4 mm.

SECTION 913
WASTE STACK VENT

913.1 Waste stack vent permitted. A waste *stack* shall be considered to be a vent for all of the fixtures discharging to the *stack* where installed in accordance with the requirements of this section.

913.2 Stack installation. The waste *stack* shall be vertical, and both horizontal and vertical offsets shall be prohibited between the lowest *fixture drain* connection and the highest *fixture drain* connection. *Fixture drains* shall connect separately to the waste *stack*. The *stack* shall not receive the discharge of water closets or urinals.

913.3 Stack vent. A *stack vent* shall be provided for the waste *stack*. The size of the *stack vent* shall be not less than the size of the waste *stack*. Offsets shall be permitted in the *stack vent*, shall be located not less than 6 inches (152 mm) above the flood level of the highest fixture and shall be in accordance with Section 905.2. The *stack vent* shall be permitted to connect with other *stack vents* and vent *stacks* in accordance with Section 904.5.

913.4 Waste stack size. The waste *stack* shall be sized based on the total discharge to the *stack* and the discharge within a *branch* interval in accordance with Table 913.4. The waste *stack* shall be the same size throughout its length.

TABLE 913.4
WASTE STACK VENT SIZE

STACK SIZE (inches)	MAXIMUM NUMBER OF DRAINAGE FIXTURE UNITS (dfu)	
	Total discharge into one branch interval	Total discharge for stack
1½	1	2
2	2	4
2½	No limit	8
3	No limit	24
4	No limit	50
5	No limit	75
6	No limit	100

For SI: 1 inch = 25.4 mm.

SECTION 914
CIRCUIT VENTING

914.1 Circuit vent permitted. Not more than eight fixtures connected to a horizontal *branch* drain shall be permitted to be circuit vented. Each *fixture drain* shall connect horizontally to the horizontal *branch* being circuit vented. The *horizontal branch drain* shall be classified as a vent from the most downstream *fixture drain* connection to the most upstream *fixture drain* connection to the horizontal *branch*.

914.1.1 Multiple circuit-vented branches. Circuit-vented horizontal *branch* drains are permitted to be connected together. Each group of not more than eight fixtures shall be considered to be a separate circuit vent and shall conform to the requirements of this section.

914.2 Vent connection. The circuit vent connection shall be located between the two most upstream *fixture drains*. The vent shall connect to the horizontal *branch* and shall be installed in accordance with Section 905. The circuit vent pipe shall not receive the discharge of any soil or waste.

914.3 Slope and size of horizontal branch. The slope of the vent section of the *horizontal branch drain* shall be not greater than one unit vertical in 12 units horizontal (8.3-percent slope). The entire length of the vent section of the *horizontal branch drain* shall be sized for the total drainage discharge to the *branch*.

914.3.1 Size of multiple circuit vent. Each separate circuit-vented horizontal *branch* that is interconnected shall be sized independently in accordance with Section 914.3. The downstream circuit-vented horizontal *branch* shall be sized for the total discharge into the *branch*, including the upstream *branches* and the fixtures within the *branch*.

914.4 Relief vent. A relief vent shall be provided for circuit-vented horizontal *branches* receiving the discharge of four or more water closets and connecting to a drainage *stack* that receives the discharge of soil or waste from upper horizontal *branches*.

914.4.1 Connection and installation. The relief vent shall connect to the horizontal *branch* drain between the *stack* and the most downstream *fixture drain* of the circuit vent. The relief vent shall be installed in accordance with Section 905.

914.4.2 Fixture drain or branch. The relief vent is permitted to be a *fixture drain* or fixture *branch* for fixtures located within the same *branch interval* as the circuit-vented horizontal *branch*. The maximum discharge to a relief vent shall be four fixture units.

914.5 Additional fixtures. Fixtures, other than the circuit-vented fixtures, are permitted to discharge to the *horizontal branch drain*. Such fixtures shall be located on the same floor as the circuit-vented fixtures and shall be either individually or common vented.

SECTION 915
COMBINATION WASTE AND VENT SYSTEM

915.1 Type of fixtures. A *combination waste and vent system* shall not serve fixtures other than floor drains, sinks, lavatories and drinking fountains. *Combination waste and vent systems* shall not receive the discharge from a food waste disposer or clinical sink.

915.2 Installation. The only vertical pipe of a *combination waste and vent system* shall be the connection between the *fixture drain* and the horizontal combination waste and vent pipe. The vertical distance shall not exceed 8 feet (2438 mm).

915.2.1 Slope. The slope of a horizontal combination waste and vent pipe shall not exceed one-half unit vertical in 12 units horizontal (4-percent slope) and shall be not less than that indicated in Table 704.1.

915.2.2 Size and length. The size of a combination waste and vent pipe shall be not less than that indicated in Table 915.2.2. The horizontal length of a *combination waste and vent system* shall be unlimited.

TABLE 915.2.2
SIZE OF COMBINATION WASTE AND VENT PIPE

DIAMETER PIPE (inches)	MAXIMUM NUMBER OF DRAINAGE FIXTURE UNITS (dfu)	
	Connecting to a horizontal branch or stack	Connecting to a building drain or building subdrain
2	3	4
2½	6	26
3	12	31
4	20	50
5	160	250
6	360	575

For SI: 1 inch = 25.4 mm.

915.2.3 Connection. The *combination waste and vent system* shall be provided with a dry vent connected at any point within the system or the system shall connect to a horizontal drain that serves vented fixtures located on the same floor. *Combination waste and vent systems* connecting to *building drains* receiving only the discharge from one or more *stacks* shall be provided with a dry vent. The vent connection to the combination waste and vent pipe shall extend vertically to a point not less than 6 inches

(152 mm) above the *flood level rim* of the highest fixture being vented before offsetting horizontally.

915.2.4 Vent size. The vent shall be sized for the total *drainage fixture unit* load in accordance with Section 906.2.

915.2.5 Fixture branch or drain. The *fixture branch* or *fixture drain* shall connect to the combination waste and vent within a distance specified in Table 909.1. The combination waste and vent pipe shall be considered to be the vent for the fixture.

SECTION 916
ISLAND FIXTURE VENTING

916.1 Limitation. Island fixture venting shall not be permitted for fixtures other than sinks and lavatories. Residential kitchen sinks with a dishwasher waste connection, a food waste disposer, or both, in combination with the kitchen sink waste, shall be permitted to be vented in accordance with this section.

916.2 Vent connection. The island fixture vent shall connect to the *fixture drain* as required for an individual or common vent. The vent shall rise vertically to above the drainage outlet of the fixture being vented before offsetting horizontally or vertically downward. The vent or *branch vent* for multiple island fixture vents shall extend to a point not less than 6 inches (152 mm) above the highest island fixture being vented before connecting to the outside vent terminal.

916.3 Vent installation below the fixture flood level rim. The vent located below the *flood level rim* of the fixture being vented shall be installed as required for drainage piping in accordance with Chapter 7, except for sizing. The vent shall be sized in accordance with Section 906.2. The lowest point of the island fixture vent shall connect full size to the drainage system. The connection shall be to a vertical drain pipe or to the top half of a horizontal drain pipe. Cleanouts shall be provided in the island fixture vent to permit rodding of all vent piping located below the *flood level rim* of the fixtures. Rodding in both directions shall be permitted through a cleanout.

SECTION 917
SINGLE-STACK VENT SYSTEM

917.1 Single-stack vent system permitted. A drainage *stack* shall serve as a single-stack vent system where sized and installed in accordance with Sections 917.2 through 917.9. The drainage *stack* and *branch* piping shall be the vents for the drainage system. The drainage *stack* shall have a *stack vent*.

917.2 Stack size. Drainage *stacks* shall be sized in accordance with Table 917.2. *Stacks* shall be uniformly sized based on the total connected *drainage fixture unit* load. The *stack vent* shall be the same size as the drainage *stack*. A 3-inch (76 mm) *stack* shall serve not more than two water closets.

TABLE 917.2
SINGLE STACK SIZE

STACK SIZE (inches)	MAXIMUM CONNECTED DRAINAGE FIXTURE UNITS		
	Stacks less than 75 feet in height	Stacks 75 feet to less than 160 feet in height	Stacks 160 feet and greater in height
3	24	NP	NP
4	225	24	NP
5	480	225	24
6	1,015	480	225
8	2,320	1,015	480
10	4,500	2,320	1,015
12	8,100	4,500	2,320
15	13,600	8,100	4,500

For SI: 1 inch = 25.4 mm, 1 foot = 304.8 mm.

917.3 Branch size. Horizontal *branches* connecting to a single-stack vent system shall be sized in accordance with Table 710.1(2). Not more than one water closet shall discharge into a 3-inch (76 mm) horizontal *branch* at a point within a developed length of 18 inches (457 mm) measured horizontally from the *stack*.

Where a water closet is within 18 inches (457 mm) measured horizontally from the *stack* and not more than one fixture with a drain size of not more than $1^1/_2$ inches (38 mm) connects to a 3-inch (76 mm) horizontal *branch*, the *branch* drain connection to the *stack* shall be made with a sanitary tee.

917.4 Length of horizontal branches. The length of horizontal *branches* shall conform to the requirements of Sections 917.4.1 through 917.4.3.

917.4.1 Water closet connection. Water closet connections shall be not greater than 4 feet (1219 mm) in *developed length* measured horizontally from the *stack*.

Exception: Where the connection is made with a sanitary tee, the maximum *developed length* shall be 8 feet (2438 mm).

917.4.2 Fixture connections. Fixtures other than water closets shall be located not greater than 12 feet (3657 mm) in developed length, measured horizontally from the *stack*.

917.4.3 Vertical piping in branch. The length of vertical piping in a *fixture drain* connecting to a horizontal *branch* shall not be considered in computing the fixture's distance in *developed length* measured horizontally from the *stack*.

917.5 Minimum vertical piping size from fixture. The vertical portion of piping in a *fixture drain* to a horizontal *branch* shall be 2 inches (51 mm). The minimum size of the vertical portion of piping for a water-supplied urinal or standpipe shall be 3 inches (76 mm). The maximum vertical drop shall be 4 feet (1219 mm). *Fixture drains* that are not increased in size, or have a vertical drop in excess of 4 feet (1219 mm), shall be individually vented.

917.6 Additional venting required. Additional venting shall be provided where more than one water closet discharges to a horizontal *branch* and where the distance from a fixture trap to the stack exceeds the limits in Section 917.4. Where additional venting is required, the fixture(s) shall be vented by individual vents, common vents, wet vents, circuit vents, or a combination waste and vent pipe. The dry vent extensions for the additional venting shall connect to a *branch vent*, vent *stack*, *stack vent*, air admittance valve, or shall terminate outdoors.

917.7 Stack offsets. Where *fixture drains* are not connected below a horizontal offset in a *stack*, a horizontal offset shall not be required to be vented. Where horizontal *branches* or *fixture drains* are connected below a horizontal offset in a *stack*, the offset shall be vented in accordance with Section 907. Fixture connections shall not be made to a *stack* within 2 feet (610 mm) above or below a horizontal offset.

917.8 Prohibited lower connections. *Stacks* greater than 2 *branch intervals* in height shall not receive the discharge of horizontal *branches* on the lower two floors. There shall not be connections to the *stack* between the lower two floors and a distance of not less than 10 pipe diameters downstream from the base of the single stack vented system.

917.9 Sizing building drains and sewers. The *building drain* and *building sewer* receiving the discharge of a single stack vent system shall be sized in accordance with Table 710.1(1).

SECTION 918
AIR ADMITTANCE VALVES

918.1 General. Vent systems utilizing air admittance valves shall comply with this section. Stack-type air admittance valves shall conform to ASSE 1050. Individual and branch-type air admittance valves shall conform to ASSE 1051.

918.2 Installation. The valves shall be installed in accordance with the requirements of this section and the manufacturer's instructions. Air admittance valves shall be installed after the DWV testing required by Section 312.2 or 312.3 has been performed.

918.3 Where permitted. Individual, branch and circuit vents shall be permitted to terminate with a connection to an individual or branch-type air admittance valve in accordance with Section 918.3.1. *Stack vents* and vent *stacks* shall be permitted to terminate to stack-type air admittance valves in accordance with Section 918.3.2.

918.3.1 Horizontal branches. Individual and branch-type air admittance valves shall vent only fixtures that are on the same floor level and connect to a *horizontal branch drain*. Where the horizontal *branch* is located more than four *branch intervals* from the top of the stack, the horizontal *branch* shall be provided with a relief vent that shall connect to a vent stack or stack vent, or extend outdoors to the open air. The relief vent shall connect to the *horizontal branch drain* between the stack and the most downstream *fixture drain* connected to the *horizontal branch drain*. The relief vent shall be sized in accordance with Section 906.2 and installed in accordance with Section 905. The relief vent shall be permitted to serve as the vent for other fixtures.

918.3.2 Stack. Stack-type air admittance valves shall be prohibited from serving as the vent terminal for vent *stacks* or *stack vents* that serve drainage *stacks* having more than six *branch intervals*.

918.4 Location. Individual and branch-type air admittance valves shall be located not less than 4 inches (102 mm) above the *horizontal branch drain* or *fixture drain* being vented. Stack-type air admittance valves shall be located not less than 6 inches (152 mm) above the *flood level rim* of the highest fixture being vented. The air admittance valve shall be located within the maximum *developed length* permitted for the vent. The air admittance valve shall be installed not less than 6 inches (152 mm) above insulation materials.

918.5 Access and ventilation. *Access* shall be provided to all air admittance valves. Such valves shall be installed in a location that allows air to enter the valve.

918.6 Size. The air admittance valve shall be rated in accordance with the standard for the size of the vent to which the valve is connected.

918.7 Vent required. Within each plumbing system, not less than one *stack vent* or vent *stack* shall extend outdoors to the open air.

918.8 Prohibited installations. Air admittance valves shall not be installed in nonneutralized special waste systems as described in Chapter 8 except where such valves are in compliance with ASSE 1049, are constructed of materials *approved* in accordance with Section 702.5 and are tested for chemical resistance in accordance with ASTM F1412. Air admittance valves shall not be located in spaces utilized as supply or return air plenums. Air admittance valves shall not be used to vent sumps or tanks except where the vent system for the sump or tank has been designed by an engineer. Air admittance valves shall not be installed on outdoor vent terminals for the sole purpose of reducing clearances to gravity air intakes or mechanical air intakes.

SECTION 919
ENGINEERED VENT SYSTEMS

919.1 General. Engineered vent systems shall comply with this section and the design, submittal, approval, inspection and testing requirements of Section 316.

919.2 Individual branch fixture and individual fixture header vents. The maximum *developed length* of individual fixture vents to vent branches and vent headers shall be determined in accordance with Table 919.2 for the minimum pipe diameters at the indicated vent airflow rates.

The individual vent airflow rate shall be determined in accordance with the following:

$$Q_{h,b} = N_{n,b} Q_v \quad \text{(Equation 9-1)}$$

For SI: $Q_{h,b} = N_{n,b} Q_v (0.4719 \text{ L/s})$

where:

$N_{n,b}$ = Number of fixtures per header (or vent *branch*) ÷ total number of fixtures connected to vent *stack*.

$Q_{h,b}$ = Vent *branch* or vent header airflow rate (cfm).

Q_v = Total vent *stack* airflow rate (cfm).

$Q_v \text{(gpm)} = 27.8 \, r_s^{2/3} (1 - r_s) D^{8/3}$

$Q_v \text{(cfm)} = 0.134 \, Q_v \text{(gpm)}$

where:

D = Drainage *stack* diameter (inches).

Q_w = Design discharge load (gpm).

r_s = Wastewater flow area to total area.

$= \dfrac{Q_w}{27.8 \, D^{8/3}}$

Individual vent airflow rates are obtained by equally distributing $Q_{h,b}$ into one-half the total number of fixtures on the *branch* or header for more than two fixtures; for an odd number of total fixtures, decrease by one; for one fixture, apply the full value of $Q_{h,b}$.

Individual vent *developed length* shall be increased by 20 percent of the distance from the vent *stack* to the fixture vent connection on the vent *branch* or header.

SECTION 920
COMPUTERIZED VENT DESIGN

920.1 Design of vent system. The sizing, design and layout of the vent system shall be permitted to be determined by *approved* computer program design methods.

920.2 System capacity. The vent system shall be based on the air capacity requirements of the drainage system under a peak load condition.

TABLE 919.2
MINIMUM DIAMETER AND MAXIMUM LENGTH OF INDIVIDUAL BRANCH FIXTURE VENTS AND INDIVIDUAL FIXTURE HEADER VENTS FOR SMOOTH PIPES

DIAMETER OF VENT PIPE (inches)	INDIVIDUAL VENT AIRFLOW RATE (cubic feet per minute)																			
	Maximum developed length of vent (feet)																			
	1	2	3	4	5	6	7	8	9	10	11	12	13	14	15	16	17	18	19	20
$^1/_2$	95	25	13	8	5	4	3	2	1	1	1	1	1	1	1	1	1	1	1	1
$^3/_4$	100	88	47	30	20	15	10	9	7	6	5	4	3	3	3	2	2	2	2	1
1	—	—	100	94	65	48	37	29	24	20	17	14	12	11	9	8	7	7	6	6
$1^1/_4$	—	—	—	—	—	—	—	100	87	73	62	53	46	40	36	32	29	26	23	21
$1^1/_2$	—	—	—	—	—	—	—	—	—	—	100	96	84	75	65	60	54	49	45
2	—	—	—	—	—	—	—	—	—	—	—	—	—	—	—	—	—	—	—	100

For SI: 1 inch = 25.4 mm, 1 cubic foot per minute = 0.4719 L/s, 1 foot = 304.8 mm.

CHAPTER 10

TRAPS, INTERCEPTORS AND SEPARATORS

User note:

About this chapter: *Chapter 10 regulates the design of fixture traps, methods for preventing evaporation of trap seals in traps and the required locations for interceptors and separators. The trap seal of a trap is an essential feature of a drainage system to prevent odors from the drainage piping from entering the building. The discharge of various processes, such as cooking and laundry, creates the need for equipment to retain detrimental greases and solids from entering the drainage systems.*

SECTION 1001
GENERAL

1001.1 Scope. This chapter shall govern the material and installation of traps, interceptors and separators.

SECTION 1002
TRAP REQUIREMENTS

1002.1 Fixture traps. Each plumbing fixture shall be separately trapped by a liquid-seal trap, except as otherwise permitted by this code. The vertical distance from the fixture outlet to the trap weir shall not exceed 24 inches (610 mm), and the horizontal distance shall not exceed 30 inches (762 mm) measured from the centerline of the fixture outlet to the centerline of the inlet of the trap. The height of a clothes washer standpipe above a trap shall conform to Section 802.3.3. A fixture shall not be double trapped.

Exceptions:

1. This section shall not apply to fixtures with integral traps.

2. A combination plumbing fixture is permitted to be installed on one trap, provided that one compartment is not more than 6 inches (152 mm) deeper than the other compartment and the waste outlets are not more than 30 inches (762 mm) apart.

3. A grease interceptor intended to serve as a fixture trap in accordance with the manufacturer's installation instructions shall be permitted to serve as the trap for a single fixture or a combination sink of not more than three compartments where the vertical distance from the fixture outlet to the inlet of the interceptor does not exceed 30 inches (762 mm) and the *developed length* of the waste pipe from the most upstream fixture outlet to the inlet of the interceptor does not exceed 60 inches (1524 mm).

4. Floor drains in multilevel parking structures that discharge to a building storm *sewer* shall not be required to be individually trapped. Where floor drains in multilevel parking structures are required to discharge to a combined *building sewer* system, the floor drains shall not be required to be individually trapped provided that they are connected to a main trap in accordance with Section 1103.1.

1002.2 Design of traps. Fixture traps shall be self-scouring. Fixture traps shall not have interior partitions, except where such traps are integral with the fixture or where such traps are constructed of an *approved* material that is resistant to corrosion and degradation. Slip joints shall be made with an *approved* elastomeric gasket and shall be installed only on the trap inlet, trap outlet and within the trap seal.

1002.3 Prohibited traps. The following types of traps are prohibited:

1. Traps that depend on moving parts to maintain the seal.
2. Bell traps.
3. Crown-vented traps.
4. Traps not integral with a fixture and that depend on interior partitions for the seal, except those traps constructed of an *approved* material that is resistant to corrosion and degradation.
5. "S" traps.
6. Drum traps.

   Exception: Drum traps used as solids interceptors and drum traps serving chemical waste systems shall not be prohibited.

1002.4 Trap seals. Each fixture trap shall have a liquid seal of not less than 2 inches (51 mm) and not more than 4 inches (102 mm), or deeper for special designs relating to accessible fixtures.

1002.4.1 Trap seal protection. Trap seals of *emergency floor drain* traps and trap seals subject to evaporation shall be protected by one of the methods in Sections 1002.4.1.1 through 1002.4.1.4.

1002.4.1.1 Potable water-supplied trap seal primer valve. A potable water-supplied trap seal primer valve shall supply water to the trap. Water-supplied trap seal primer valves shall conform to ASSE 1018. The discharge pipe from the trap seal primer valve shall connect to the trap above the trap seal on the inlet side of the trap.

1002.4.1.2 Reclaimed water or graywater-supplied trap seal primer valve. A reclaimed water or graywater-supplied trap seal primer valve shall supply water to the trap. Water-supplied trap seal primer valves shall conform to ASSE 1018. The quality of reclaimed water or graywater-supplied to trap seal primer valves shall be in accordance with the requirements of the manufacturer of the trap seal primer valve. The discharge pipe from the trap seal primer valve shall connect to the trap above the trap seal, on the inlet side of the trap.

1002.4.1.3 Wastewater-supplied trap primer device. A wastewater-supplied trap primer device shall supply water to the trap. Wastewater-supplied trap primer devices shall conform to ASSE 1044. The discharge pipe from the trap seal primer device shall connect to the trap above the trap seal on the inlet side of the trap.

1002.4.1.4 Barrier-type trap seal protection device. A barrier-type trap seal protection device shall protect the floor drain trap seal from evaporation. Barrier-type floor drain trap seal protection devices shall conform to ASSE 1072. The devices shall be installed in accordance with the manufacturer's instructions.

1002.5 Size of fixture traps. Fixture trap size shall be sufficient to drain the fixture rapidly and not less than the size indicated in Table 709.1. A trap shall not be larger than the drainage pipe into which the trap discharges.

1002.6 Building traps. Building (house) traps shall be prohibited.

1002.7 Trap setting and protection. Traps shall be set level with respect to the trap seal and, where necessary, shall be protected from freezing.

1002.8 Recess for trap connection. A recess provided for connection of the underground trap, such as one serving a bathtub in slab-type construction, shall have sides and a bottom of corrosion-resistant, insect- and verminproof construction.

1002.9 Acid-resisting traps. Where a vitrified clay or other brittleware, acid-resisting trap is installed underground, such trap shall be embedded in concrete extending 6 inches (152 mm) beyond the bottom and sides of the trap.

1002.10 Plumbing in mental health centers. In mental health centers, pipes and traps shall not be exposed.

SECTION 1003
INTERCEPTORS AND SEPARATORS

1003.1 Where required. Interceptors and separators shall be provided to prevent the discharge of oil, grease, sand and other substances harmful or hazardous to the public sewer, the private sewage system or the sewage treatment plant or processes.

1003.2 Approval. The size, type and location of each interceptor and of each separator shall be designed and installed in accordance with the manufacturer's instructions and the requirements of this section based on the anticipated conditions of use. Wastes that do not require treatment or separation shall not be discharged into any interceptor or separator.

1003.3 Grease interceptors. Grease interceptors shall comply with the requirements of Sections 1003.3.1 through 1003.3.8.

1003.3.1 Grease interceptors and automatic grease removal devices required. A grease interceptor or automatic grease removal device shall be required to receive the drainage from fixtures and equipment with grease-laden waste located in food preparation areas, such as in restaurants, hotel kitchens, hospitals, school kitchens, bars, factory cafeterias and clubs. Fixtures and equipment shall include pot sinks, prerinse sinks; soup kettles or similar devices; wok stations; floor drains or sinks into which kettles are drained; automatic hood wash units and dishwashers without prerinse sinks. Grease interceptors and automatic grease removal devices shall receive waste only from fixtures and equipment that allow fats, oils or grease to be discharged. Where lack of space or other constraints prevent the installation or replacement of a grease interceptor, one or more grease interceptors shall be permitted to be installed on or above the floor and upstream of an existing grease interceptor.

1003.3.2 Food waste disposers restriction. A food waste disposer shall not discharge to a grease interceptor.

1003.3.3 Additives to grease interceptors. Dispensing systems that dispense interceptor performance additives to grease interceptors shall not be installed except where such systems dispense microbes for the enhancement of aerobic bioremediation of grease and other organic material, or for inhibiting growth of pathogenic organisms by anaerobic methods. Such microbial dispensing systems shall be installed only where the grease interceptor manufacturer's instructions allow such systems and the systems conform to ASME A112.14.6. Systems that discharge emulsifiers, chemicals or enzymes to grease interceptors shall be prohibited.

1003.3.4 Grease interceptors and automatic grease removal devices not required. A grease interceptor or an automatic grease removal device shall not be required for individual dwelling units or any private living quarters.

1003.3.5 Hydromechanical grease interceptors, fats, oils and greases disposal systems and automatic grease removal devices. Hydromechanical grease interceptors; fats, oils, and greases disposal systems and automatic grease removal devices shall be sized in accordance with ASME A112.14.3, ASME A112.14.4, ASME A112.14.6, CSA B481.3 or PDI G101. Hydromechanical grease interceptors; fats, oils, and greases disposal systems and automatic grease removal devices shall be designed and tested in accordance with ASME A112.14.3, ASME A112.14.4, CSA B481.1, PDI G101 or PDI G102. Hydromechanical grease interceptors; fats, oils, and greases disposal systems and automatic grease removal devices shall be installed in accordance with the manufacturer's instructions. Where manufacturer's instructions are not provided, hydromechanical grease interceptors; fats, oils, and greases disposal systems and automatic grease removal devices shall be installed in compliance with ASME A112.14.3, ASME A112.14.4, ASME A112.14.6, CSA B481.3 or PDI G101.

1003.3.5.1 Grease interceptor capacity. Grease interceptors shall have the grease retention capacity indicated in Table 1003.3.5.1 for the flow-through rates indicated.

TRAPS, INTERCEPTORS AND SEPARATORS

TABLE 1003.3.5.1
CAPACITY OF GREASE INTERCEPTORS[a]

TOTAL FLOW-THROUGH RATING (gpm)	GREASE RETENTION CAPACITY (pounds)
4	8
6	12
7	14
9	18
10	20
12	24
14	28
15	30
18	36
20	40
25	50
35	70
50	100
75	150
100	200

For SI: 1 gallon per minute = 3.785 L/m, 1 pound = 0.454 kg.

a. For total flow-through ratings greater than 100 (gpm), double the flow-through rating to determine the grease retention capacity (pounds).

1003.3.5.2 Rate of flow controls. Grease interceptors shall be equipped with devices to control the rate of water flow so that the water flow does not exceed the rated flow. The flow-control device shall be vented and terminate not less than 6 inches (152 mm) above the flood rim level or be installed in accordance with the manufacturer's instructions.

1003.3.6 Automatic grease removal devices. Where automatic grease removal devices are installed, such devices shall be located downstream of each fixture or multiple fixtures in accordance with the manufacturer's instructions. The automatic grease removal device shall be sized to pretreat the measured or calculated flows for all connected fixtures or equipment. Ready *access* shall be provided for inspection and maintenance.

1003.3.7 Gravity grease interceptors and gravity grease interceptors with fats, oils, and greases disposal systems. The required capacity of gravity grease interceptors and gravity grease interceptors with fats, oils, and greases disposal systems shall be determined by multiplying the peak drain flow into the interceptor in gallons per minute by a retention time of 30 minutes. Gravity grease interceptors shall be designed and tested in accordance with IAPMO/ANSI Z1001. Gravity grease interceptors with fats, oils, and greases disposal systems shall be designed and tested in accordance with ASME A112.14.6 and IAPMO/ANSI Z1001. Gravity grease interceptors and gravity grease interceptors with fats, oils, and greases disposal systems shall be installed in accordance with manufacturer's instructions. Where manufacturer's instructions are not provided, gravity grease interceptors and gravity grease interceptors with fats, oils, and greases disposal systems shall be installed in compliance with ASME A112.14.6 and IAPMO/ANSI Z1001.

1003.3.8 Direct connection. The discharge piping from a grease interceptor shall be directly connected to the sanitary drainage system.

1003.4 Oil separators required. At repair garages where floor or trench drains are provided, car washing facilities, factories where oily and flammable liquid wastes are produced and hydraulic elevator pits, oil separators shall be installed into which oil-bearing, grease-bearing or flammable wastes shall be discharged before emptying into the building drainage system or other point of disposal.

Exception: An oil separator is not required in hydraulic elevator pits where an approved alarm system is installed. Such alarm systems shall not terminate the operation of pumps utilized to maintain emergency operation of the elevator by fire fighters.

1003.4.1 Separation of liquids. A mixture of treated or untreated light and heavy liquids with various specific gravities shall be separated in an *approved* receptacle.

1003.4.2 Oil separator design. Oil separators shall be listed and labeled, or designed in accordance with Sections 1003.4.2.1 and 1003.4.2.2.

1003.4.2.1 General design requirements. Oil separators shall have a depth of not less than 2 feet (610 mm) below the invert of the discharge drain. The outlet opening of the separator shall have not less than an 18-inch (457 mm) water seal.

1003.4.2.2 Garages and service stations. Where automobiles are serviced, greased, repaired or washed or where gasoline is dispensed, oil separators shall have a capacity of not less than 6 cubic feet (0.168 m^3) for the first 100 square feet (9.3 m^2) of area to be drained, plus 1 cubic foot (0.028 m^3) for each additional 100 square feet (9.3 m^2) of area to be drained into the separator. Parking garages in which servicing, repairing or washing is not conducted, and in which gasoline is not dispensed, shall not require a separator. Areas of commercial garages utilized only for storage of automobiles are not required to be drained through a separator.

1003.5 Sand interceptors in commercial establishments. Sand and similar interceptors for heavy solids shall be designed and located so as to be provided with ready *access* for cleaning, and shall have a water seal of not less than 6 inches (152 mm).

1003.6 Clothes washer discharge interceptor. Clothes washers shall discharge through an interceptor that is provided with a wire basket or similar device, removable for cleaning, that prevents passage into the drainage system of solids $^1/_2$ inch (12.7 mm) or larger in size, string, rags, buttons or other materials detrimental to the public sewage system.

Exceptions:

1. Clothes washers in individual dwelling units shall not be required to discharge through an interceptor.

2. A single clothes washer designed for use in individual dwelling units and installed in a location other than an individual dwelling unit shall not be required to discharge through an interceptor.

1003.7 Bottling establishments. Bottling plants shall discharge process wastes into an interceptor that will provide for the separation of broken glass or other solids before discharging waste into the drainage system.

1003.8 Slaughterhouses. Slaughtering room and dressing room drains shall be equipped with *approved* separators. The separator shall prevent the discharge into the drainage system of feathers, entrails and other materials that cause clogging.

1003.9 Venting of interceptors and separators. Interceptors and separators shall be designed so as not to become air bound. Interceptors and separators shall be vented in accordance with one of the methods in Chapter 9.

1003.10 Access and maintenance of interceptors and separators. *Access* shall be provided to each interceptor and separator for service and maintenance. Interceptors and separators shall be maintained by periodic removal of accumulated grease, scum, oil, or other floating substances and solids deposited in the interceptor or separator.

SECTION 1004
MATERIALS, JOINTS AND CONNECTIONS

1004.1 General. The materials and methods utilized for the construction and installation of traps, interceptors and separators shall comply with this chapter and the applicable provisions of Chapters 4 and 7. The fittings shall not have ledges, shoulders or reductions capable of retarding or obstructing flow of the piping.

CHAPTER 11

STORM DRAINAGE

> **User note:**
>
> *About this chapter:* Rainfall onto buildings must be removed and directed to a location that can accommodate storm water. Chapter 11 specifies the design rainfall event for the geographic area and provides sizing methods for piping and gutter systems to convey the storm water away from the building. Included in this chapter are regulations for piping materials and subsoil drainage systems.

SECTION 1101
GENERAL

1101.1 Scope. The provisions of this chapter shall govern the materials, design, construction and installation of storm drainage.

1101.2 Disposal. Rainwater from roofs and storm water from paved areas, yards, courts and courtyards shall drain to an *approved* place of disposal. For one- and two-family dwellings, and where *approved*, storm water is permitted to discharge onto flat areas, such as streets or lawns, provided that the storm water flows away from the building.

1101.3 Prohibited drainage. Storm water shall not be drained into *sewers* intended for sewage only.

1101.4 Tests. The conductors and the building *storm drain* shall be tested in accordance with Section 312.

1101.5 Change in size. The size of a drainage pipe shall not be reduced in the direction of flow.

1101.6 Fittings and connections. Connections and changes in direction of the storm drainage system shall be made with *approved* drainage-type fittings in accordance with Table 706.3. The fittings shall not obstruct or retard flow in the system.

[BS] 1101.7 Roof design. Roofs shall be designed for the maximum possible depth of water that will pond thereon as determined by the relative levels of roof deck and overflow weirs, scuppers, edges or serviceable drains in combination with the deflected structural elements. In determining the maximum possible depth of water, all primary roof drainage means shall be assumed to be blocked. The maximum possible depth of water on the roof shall include the height of the water required above the inlet of the secondary roof drainage means to achieve the required flow rate of the secondary drainage means to accommodate the design rainfall rate as required by Section 1106.

1101.8 Cleanouts required. Cleanouts shall be installed in the storm drainage system and shall comply with the provisions of this code for sanitary drainage pipe cleanouts.

> Exception: Subsurface drainage system.

1101.9 Backwater valves. Storm drainage systems shall be provided with backwater valves as required for sanitary drainage systems in accordance with Section 714.

SECTION 1102
MATERIALS

1102.1 General. The materials and methods utilized for the construction and installation of storm drainage systems shall comply with this section and the applicable provisions of Chapter 7.

1102.2 Inside storm drainage conductors. Inside storm drainage conductors installed above ground shall conform to one of the standards listed in Table 702.1.

1102.3 Underground building storm drain pipe. Underground building *storm drain* pipe shall conform to one of the standards listed in Table 702.2.

1102.4 Building storm sewer pipe. Building storm *sewer* pipe shall conform to one of the standards listed in Table 1102.4.

TABLE 1102.4
BUILDING STORM SEWER PIPE

MATERIAL	STANDARD
Acrylonitrile butadiene styrene (ABS) plastic pipe in IPS diameters, including Schedule 40, DR 22 (PS 200) and DR 24 (PS 140); with a solid, cellular core or composite wall.	ASTM D2661; ASTM F628; ASTM F1488; CSA B181.1; CSA B182.1
Cast-iron pipe	ASTM A74; ASTM A888; CISPI 301
Concrete pipe	ASTM C14; ASTM C76; CSA A257.1M; CSA A257.2M
Copper or copper-alloy tubing (Type K, L, M or DWV)	ASTM B75; ASTM B88; ASTM B251; ASTM B306
Polyethylene (PE) plastic pipe	ASTM F667; ASTM F2306/F2306M; ASTM F2648/F2648M
Polypropylene (PP) pipe	ASTM F2881; CSA B182.13
Polyvinyl chloride (PVC) plastic pipe (Type DWV, SDR26, SDR35, SDR41, PS50 or PS100) in IPS diameters, including Schedule 40, DR 22 (PS 200) and DR 24 (PS 140); with a solid, cellular core or composite wall.	ASTM D2665; ASTM D3034; ASTM F891; ASTM F1488; CSA B182.4; CSA B181.2; CSA B182.2
Vitrified clay pipe	ASTM C4; ASTM C700
Stainless steel drainage systems, Type 316L	ASME A112.3.1

STORM DRAINAGE

1102.5 Subsoil drain pipe. Subsoil drains shall be open-jointed, horizontally split or perforated pipe conforming to one of the standards listed in Table 1102.5.

1102.6 Roof Drains. Roof drains shall conform to ASME A112.6.4 or ASME A112.3.1.

TABLE 1102.5
SUBSOIL DRAIN PIPE

MATERIAL	STANDARD
Cast-iron pipe	ASTM A74; ASTM A888; CISPI 301
Polyethylene (PE) plastic pipe	ASTM F405; ASTM F667; CSA B182.1; CSA B182.6; CSA B182.8
Polyvinyl chloride (PVC) Plastic pipe (type sewer pipe, SDR35, PS25, PS50 or PS100)	ASTM D2729; ASTM D3034, ASTM F891; CSA B182.2; CSA B182.4
Stainless steel drainage systems, Type 316L	ASME A112.3.1
Vitrified clay pipe	ASTM C4; ASTM C700

1102.7 Fittings. Pipe fittings shall be *approved* for installation with the piping material installed, and shall conform to the respective pipe standards or one of the standards listed in Table 1102.7. The fittings shall not have ledges, shoulders or reductions capable of retarding or obstructing flow in the piping. Threaded drainage pipe fittings shall be of the recessed drainage type.

TABLE 1102.7
PIPE FITTINGS

MATERIAL	STANDARD
Acrylonitrile butadiene styrene (ABS) plastic	ASTM D2661; ASTM D3311; CSA B181.1
Cast-iron	ASME B16.4; ASME B16.12; ASTM A888; CISPI 301; ASTM A74
Coextruded composite ABS and drain DR-PS in PS35, PS50, PS100, PS140, PS200	ASTM D2751
Coextruded composite ABS DWV Schedule 40 IPS pipe (solid or cellular core)	ASTM D2661; ASTM D3311; ASTM F628
Coextruded composite PVC DWV Schedule 40 IPS-DR, PS140, PS200 (solid or cellular core)	ASTM D2665; ASTM D3311; ASTM F891
Coextruded composite PVC sewer and drain DR-PS in PS35, PS50, PS100, PS140, PS200	ASTM D3034
Copper or copper alloy	ASME B16.15; ASME B16.18; ASME B16.22; ASME B16.23; ASME B16.26; ASME B16.29
Gray iron and ductile iron	AWWA C110/A21.10
Malleable iron	ASME B16.3
Plastic, general	ASTM F409
Polyethylene (PE) plastic pipe	ASTM F2306/F2306M
Polyvinyl chloride (PVC) plastic	ASTM D2665; ASTM D3311; ASTM F1866
Steel	ASME B16.9; ASME B16.11; ASME B16.28
Stainless steel drainage systems, Type 316L	ASME A112.3.1

SECTION 1103 TRAPS

1103.1 Main trap. Leaders and *storm drains* connected to a combined sewer shall be trapped. Individual storm water traps shall be installed on the storm water drain *branch* serving each conductor, or a single trap shall be installed in the main *storm drain* just before its connection with the combined *building sewer* or the *public sewer*. Leaders and storm drains connected to a building storm sewer shall not be required to be trapped.

1103.2 Material. Storm water traps shall be of the same material as the piping system to which they are attached.

1103.3 Size. Traps for individual conductors shall be the same size as the horizontal drain to which they are connected.

1103.4 Cleanout. A cleanout shall be installed on the building side of the trap and shall be provided with *access*.

SECTION 1104 CONDUCTORS AND CONNECTIONS

1104.1 Prohibited use. Conductor pipes shall not be used as soil, waste or vent pipes, and soil, waste or vent pipes shall not be used as conductors.

1104.2 Floor drains. Floor drains shall not be connected to a *storm drain*.

SECTION 1105 ROOF DRAINS

1105.1 General. Roof drains shall be installed in accordance with the manufacturer's instructions. The inside opening for the roof drain shall not be obstructed by the roofing membrane material.

1105.2 Roof drain flow rate. The published roof drain flow rate, based on the head of water above the roof drain, shall be used to size the storm drainage system in accordance with Section 1106. The flow rate used for sizing the storm drainage piping shall be based on the maximum anticipated ponding at the roof drain.

SECTION 1106 SIZE OF CONDUCTORS, LEADERS AND STORM DRAINS

1106.1 General. The size of the vertical conductors and leaders, building *storm drains*, building storm *sewers* and any horizontal branches of such drains or *sewers* shall be based on the 100-year hourly rainfall rate indicated in Figure 1106.1 or on other rainfall rates determined from *approved* local weather data.

STORM DRAINAGE

For SI: 1 inch = 25.4 mm.
Source: National Weather Service, National Oceanic and Atmospheric Administration, Washington D.C.

**FIGURE 1106.1
100-YEAR, 1-HOUR RAINFALL (INCHES) EASTERN UNITED STATES**

STORM DRAINAGE

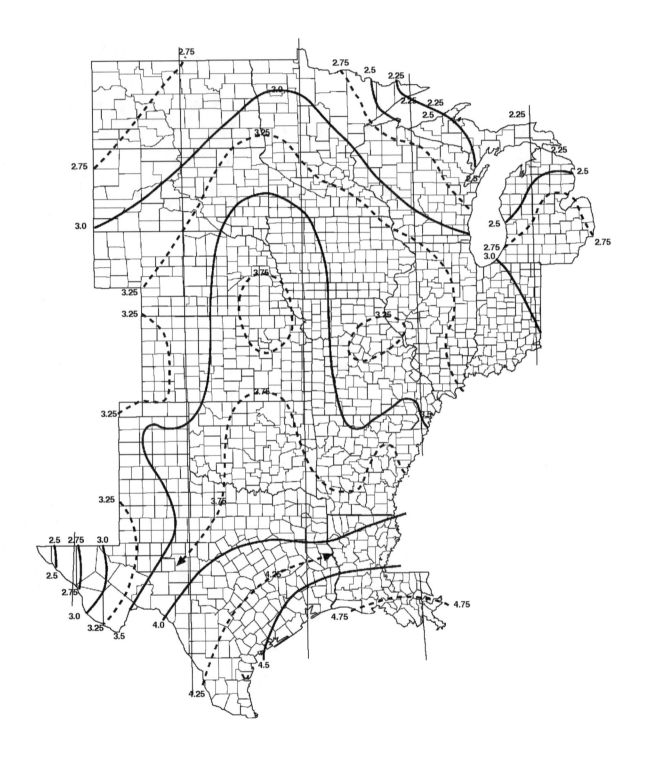

For SI: 1 inch = 25.4 mm.
Source: National Weather Service, National Oceanic and Atmospheric Administration, Washington D.C.

FIGURE 1106.1—continued
100-YEAR, 1-HOUR RAINFALL (INCHES) CENTRAL UNITED STATES

For SI: 1 inch = 25.4 mm.
Source: National Weather Service, National Oceanic and Atmospheric Administration, Washington D.C.

FIGURE 1106.1—continued
100-YEAR, 1-HOUR RAINFALL (INCHES) WESTERN UNITED STATES

STORM DRAINAGE

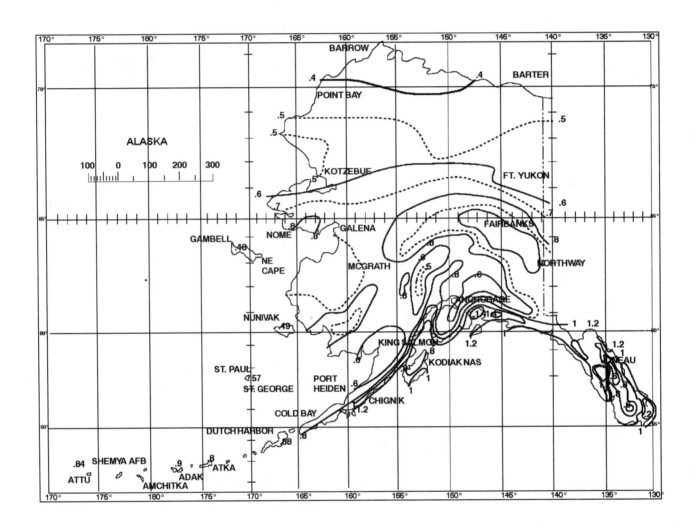

For SI: 1 inch = 25.4 mm.
Source: National Weather Service, National Oceanic and Atmospheric Administration, Washington D.C.

FIGURE 1106.1—continued
100-YEAR, 1-HOUR RAINFALL (INCHES) ALASKA

STORM DRAINAGE

For SI: 1 inch = 25.4 mm.
Source: National Weather Service, National Oceanic and Atmospheric Administration, Washington D.C.

FIGURE 1106.1—continued
100-YEAR, 1-HOUR RAINFALL (INCHES) HAWAII

STORM DRAINAGE

1106.2 Size of storm drain piping. Vertical and horizontal storm drain piping shall be sized based on the flow rate through the roof drain. The flow rate in storm drain piping shall not exceed that specified in Table 1106.2.

1106.3 Vertical leader sizing. Vertical leaders shall be sized based on the flow rate from horizontal gutters or the maximum flow rate through roof drains. The flow rate through vertical leaders shall not exceed that specified in Table 1106.3.

1106.4 Vertical walls. In sizing roof drains and storm drainage piping, one-half of the area of any vertical wall that diverts rainwater to the roof shall be added to the projected roof area for inclusion in calculating the required size of vertical conductors, leaders and horizontal storm drainage piping.

1106.5 Parapet wall scuppers. Where scuppers are used for primary roof drainage or for secondary (emergency overflow) roof drainage or both, the quantity, size, location and inlet elevation of the scuppers shall be chosen to prevent the depth of ponding water on the roof from exceeding the maximum water depth that the roof was designed for as determined by Section 1611.1 of the *International Building Code*. Scupper openings shall be not less than 4 inches (102 mm) in height and have a width that is equal to or greater than the circumference of a roof drain sized for the same roof area. The flow through the primary system shall not be considered when locating and sizing secondary scuppers.

1106.6 Size of roof gutters. Horizontal gutters shall be sized based on the flow rate from the roof surface. The flow rate in horizontal gutters shall not exceed that specified in Table 1106.6.

SECTION 1107
SIPHONIC ROOF DRAINAGE SYSTEMS

1107.1 General. Siphonic roof drains and drainage systems shall be designed in accordance with ASME A112.6.9 and ASPE 45.

SECTION 1108
SECONDARY (EMERGENCY) ROOF DRAINS

1108.1 Secondary (emergency overflow) drains or scuppers. Where roof drains are required, secondary (emergency overflow) roof drains or scuppers shall be provided where the roof perimeter construction extends above the roof in such a manner that water will be entrapped if the primary drains allow buildup for any reason. Where primary and secondary roof drains are manufactured as a single assembly, the inlet and outlet for each drain shall be independent.

TABLE 1106.3
VERTICAL LEADER SIZING

SIZE OF LEADER (inches)	CAPACITY (gpm)
2	30
2 × 2	30
1$\frac{1}{2}$ × 2$\frac{1}{2}$	30
2$\frac{1}{2}$	54
2$\frac{1}{2}$ × 2$\frac{1}{2}$	54
3	92
2 × 4	92
2$\frac{1}{2}$ × 3	92
4	192
3 × 4$\frac{1}{4}$	192
3$\frac{1}{2}$ × 4	192
5	360
4 × 5	360
4$\frac{1}{2}$ × 4$\frac{1}{2}$	360
6	563
5 × 6	563
5$\frac{1}{2}$ × 5$\frac{1}{2}$	563
8	1208
6 × 8	1208

For SI: 1 inch = 25.4 mm, 1 gallon per minute = 3.785 L/m.

TABLE 1106.2
STORM DRAIN PIPE SIZING

PIPE SIZE (inches)	CAPACITY (gpm)				
	VERTICAL DRAIN	SLOPE OF HORIZONTAL DRAIN			
		$\frac{1}{16}$ inch per foot	$\frac{1}{8}$ inch per foot	$\frac{1}{4}$ inch per foot	$\frac{1}{2}$ inch per foot
2	34	15	22	31	44
3	87	39	55	79	111
4	180	81	115	163	231
5	311	117	165	234	331
6	538	243	344	487	689
8	1,117	505	714	1,010	1,429
10	2,050	927	1,311	1,855	2,623
12	3,272	1,480	2,093	2,960	4,187
15	5,543	2,508	3,546	5,016	7,093

For SI: 1 inch = 25.4 mm, 1 foot = 304.8 mm, 1 gallon per minute = 3.785 L/m.

1108.2 Separate systems required. Secondary roof drain systems shall have the end point of discharge separate from the primary system. Discharge shall be above grade, in a location that would normally be observed by the building occupants or maintenance personnel.

1108.3 Sizing of secondary drains. Secondary (emergency) roof drain systems shall be sized in accordance with Section 1106 based on the rainfall rate for which the primary system is sized. Scuppers shall be sized to prevent the depth of ponding water from exceeding that for which the roof was designed as determined by Section 1101.7. Scuppers shall have an opening dimension of not less than 4 inches (102 mm) in height and have an opening width equal to the circumference of the roof drain required for the area served. The flow through the primary system shall not be considered when sizing the secondary roof drain system.

SECTION 1109
COMBINED SANITARY AND STORM PUBLIC SEWER

1109.1 General. Where the *public sewer* is a combined system for both sanitary and storm water, the *storm sewer* shall be connected independently to the *public sewer*.

SECTION 1110
CONTROLLED FLOW ROOF DRAIN SYSTEMS

1110.1 General. The roof of a structure shall be designed for the storage of water where the storm drainage system is engineered for controlled flow. The controlled flow roof drain system shall be an engineered system in accordance with this section and the design, submittal, approval, inspection and testing requirements of Section 316.1. The controlled flow system shall be designed based on the required rainfall rate in accordance with Section 1106.1.

1110.2 Control devices. The control devices shall be installed so that the rate of discharge of water per minute shall not exceed the values for continuous flow as indicated in Section 1110.1.

1110.3 Installation. Runoff control shall be by control devices. Control devices shall be protected by strainers.

1110.4 Minimum number of roof drains. Not less than two roof drains shall be installed in roof areas 10,000 square feet (929 m^2) or less and not less than four roof drains shall be installed in roofs over 10,000 square feet (929 m^2) in area.

SECTION 1111
SUBSOIL DRAINS

1111.1 Subsoil drains. Subsoil drains shall be open-jointed, horizontally split or perforated pipe conforming to one of the standards listed in Table 1102.5. Such drains shall be not less than 4 inches (102 mm) in diameter. Where the building is subject to backwater, the subsoil drain shall be protected by an accessibly located backwater valve. Subsoil drains shall discharge to a trapped area drain, sump, dry well or *approved* location above ground. The subsoil sump shall not be required to have either a gas-tight cover or a vent. The sump and pumping system shall comply with Section 1113.1.

TABLE 1106.6
HORIZONTAL GUTTER SIZING

GUTTER DIMENSIONS[a] (inches)	SLOPE (inch per foot)	CAPACITY (gpm)
1^1/$_2$ × 2^1/$_2$	^{1}/$_4$	26
1^1/$_2$ × 2^1/$_2$	^{1}/$_2$	40
4	^{1}/$_8$	39
2^1/$_4$ × 3	^{1}/$_4$	55
2^1/$_4$ × 3	^{1}/$_2$	87
5	^{1}/$_8$	74
4 × 2^1/$_2$	^{1}/$_4$	106
3 × 3^1/$_2$	^{1}/$_2$	156
6	^{1}/$_8$	110
3 × 5	^{1}/$_4$	157
3 × 5	^{1}/$_2$	225
8	^{1}/$_{16}$	172
8	^{1}/$_8$	247
4^1/$_2$ × 6	^{1}/$_4$	348
4^1/$_2$ × 6	^{1}/$_2$	494
10	^{1}/$_{16}$	331
10	^{1}/$_8$	472
5 × 8	^{1}/$_4$	651
4 × 10	^{1}/$_2$	1055

For SI: 1 inch = 25.4 mm, 1 foot = 304.8 mm, 1 gallon per minute = 3.785 L/m, 1 inch per foot = 83.3 mm/m.

a. Dimensions are width by depth for rectangular shapes. Single dimensions are diameters of a semicircle.

SECTION 1112
BUILDING SUBDRAINS

1112.1 Building subdrains. *Building subdrains* located below the *public sewer* level shall discharge into a sump or receiving tank, the contents of which shall be automatically lifted and discharged into the drainage system as required for building sumps. The sump and pumping equipment shall comply with Section 1113.1.

SECTION 1113
SUMPS AND PUMPING SYSTEMS

1113.1 Pumping system. The sump pump, pit and discharge piping shall conform to Sections 1113.1.1 through 1113.1.4.

1113.1.1 Pump capacity and head. The sump pump shall be of a capacity and head appropriate to anticipated use requirements.

1113.1.2 Sump pit. The sump pit shall be not less than 18 inches (457 mm) in diameter and not less than 24 inches (610 mm) in depth, unless otherwise *approved*. The pit shall be provided with access and shall be located such that all drainage flows into the pit by gravity. The sump pit shall be constructed of tile, steel, plastic, cast iron, concrete or other *approved* material, with a removable cover adequate to support anticipated loads in the area of use. The pit floor shall be solid and provide permanent support for the pump.

1113.1.3 Electrical. Electrical service outlets, where required, shall meet the requirements of NFPA 70.

1113.1.4 Piping. Discharge piping shall meet the requirements of Section 1102.2, 1102.3 or 1102.4 and shall include a gate valve and a full flow check valve. Pipe and fittings shall be the same size as, or larger than, the pump discharge tapping.

> **Exception:** In one- and two-family dwellings, only a check valve shall be required, located on the discharge piping from the pump or ejector.

CHAPTER 12

SPECIAL PIPING AND STORAGE SYSTEMS

User note:

About this chapter: Chapter 12 specifies the standards covering the installation of nonflammable medical gas piping systems and nonmedical oxygen piping systems.

SECTION 1201
GENERAL

1201.1 Scope. The provisions of this chapter shall govern the design and installation of piping and storage systems for nonflammable medical gas systems and nonmedical oxygen systems. All maintenance and operations of such systems shall be in accordance with the *International Fire Code*.

SECTION 1202
MEDICAL GASES

[F] 1202.1 Nonflammable medical gases. Nonflammable medical gas systems, inhalation anesthetic systems and vacuum piping systems shall be designed and installed in accordance with NFPA 99.

Exceptions:

1. This section shall not apply to portable systems or cylinder storage.

2. Vacuum system exhaust terminations shall comply with the *International Mechanical Code*.

SECTION 1203
OXYGEN SYSTEMS

[F] 1203.1 Design and installation. Nonmedical oxygen systems shall be designed and installed in accordance with NFPA 55 and NFPA 51.

CHAPTER 13

NONPOTABLE WATER SYSTEMS

> **User note:**
> *About this chapter: Storm water and some liquid waste from a building can be a source of nonpotable water that can be used to reduce the volume of potable water supplied to the building. Chapter 13 provides the requirements for storage, treatment and distribution of this resource. This chapter also regulates the piping systems for reclaimed water supplied by a wastewater treatment facility.*

SECTION 1301
GENERAL

1301.1 Scope. The provisions of Chapter 13 shall govern the materials, design, construction and installation of systems for the collection, storage, treatment and distribution of nonpotable water. The use and application of nonpotable water shall comply with laws, rules and ordinances applicable in the jurisdiction.

1301.2 Water quality. Nonpotable water for each end use application shall meet the minimum water quality requirements as established for the intended application by the laws, rules and ordinances applicable in the jurisdiction. Where nonpotable water from different sources is combined in a system, the system shall comply with the most stringent of the requirements of this code that are applicable to such sources.

1301.2.1 Residual disinfectants. Where chlorine is used for disinfection, the nonpotable water shall contain not more than 4 ppm (4mg/L) of chloramines or free chlorine when tested in accordance with ASTM D1253. Where ozone is used for disinfection, the nonpotable water shall not contain gas bubbles having elevated levels of ozone at the point of use.

> **Exception:** Reclaimed water sources shall not be required to comply with these requirements.

1301.2.2 Filtration required. Nonpotable water utilized for water closet and urinal flushing applications shall be filtered by a 100-micron or finer filter.

> **Exception:** Reclaimed water sources shall not be required to comply with these requirements.

1301.3 Signage required. Nonpotable water outlets such as hose connections, open ended pipes and faucets shall be identified at the point of use for each outlet with signage that reads as follows: "Nonpotable water is utilized for [application name]. CAUTION: NONPOTABLE WATER – DO NOT DRINK." The words shall be legibly and indelibly printed on a tag or sign constructed of corrosion-resistant waterproof material or shall be indelibly printed on the fixture. The letters of the words shall be not less than 0.5 inch (12.7 mm) in height and in colors in contrast to the background on which they are applied. In addition to the required wordage, the pictograph shown in Figure 1301.3 shall appear on the signage required by this section.

1301.4 Permits. Permits shall be required for the construction, installation, alteration and repair of nonpotable water systems. Construction documents, engineering calculations, diagrams and other such data pertaining to the nonpotable water system shall be submitted with each permit application.

1301.5 Potable water connections. Where a potable system is connected to a nonpotable water system, the potable water supply shall be protected against backflow in accordance with Section 608.

1301.6 Components and materials. Piping, plumbing components and materials used in collection and conveyance systems shall be of material approved by the manufacturer for the intended application.

1301.7 Insect and vermin control. The system shall be protected to prevent the entrance of insects and vermin into storage tanks and piping systems. Screen materials shall be compatible with contacting system components and shall not accelerate the corrosion of system components.

1301.8 Freeze protection. Where sustained freezing temperatures occur, provisions shall be made to keep storage tanks and the related piping from freezing.

1301.9 Nonpotable water storage tanks. Nonpotable water storage tanks shall comply with Sections 1301.9.1 through 1301.9.10.

**FIGURE 1301.3
PICTOGRAPH—DO NOT DRINK**

1301.9.1 Location. Any storage tank or portion thereof that is above grade shall be protected from direct exposure to sunlight by one of the following methods:

1. Tank construction using opaque, UV-resistant materials such as heavily tinted plastic, fiberglass, lined metal, concrete, wood, or painted to prevent algae growth.
2. Specially constructed sun barriers.
3. Installation in garages, crawl spaces or sheds.

1301.9.2 Materials. Where collected on site, water shall be collected in an *approved* tank constructed of durable, nonabsorbent and corrosion-resistant materials. The storage tank shall be constructed of materials compatible with any disinfection systems used to treat water upstream of the tank and with any systems used to maintain water quality in the tank. Wooden storage tanks that are not equipped with a makeup water source shall be provided with a flexible liner.

1301.9.3 Foundation and supports. Storage tanks shall be supported on a firm base capable of withstanding the weight of the storage tank when filled to capacity. Storage tanks shall be supported in accordance with the *International Building Code*.

1301.9.3.1 Ballast. Where the soil can become saturated, an underground storage tank shall be ballasted, or otherwise secured, to prevent the tank from floating out of the ground when empty. The combined weight of the tank and hold down ballast shall meet or exceed the buoyancy force of the tank. Where the installation requires a foundation, the foundation shall be flat and shall be designed to support the weight of the storage tank when full, consistent with the bearing capability of adjacent soil.

1301.9.3.2 Structural support. Where installed below grade, storage tank installations shall be designed to withstand earth and surface structural loads without damage and with minimal deformation when empty or filled with water.

1301.9.4 Makeup water. Where an uninterrupted supply is required for the intended application, potable or reclaimed water shall be provided as a source of makeup water for the storage tank. The makeup water supply shall be protected against backflow in accordance with Section 608. A *full-open valve* located on the makeup water supply line to the storage tank shall be provided. Inlets to the storage tank shall be controlled by fill valves or other automatic supply valves installed to prevent the tank from overflowing and to prevent the water level from dropping below a predetermined point. Where makeup water is provided, the water level shall not be permitted to drop below the source water inlet or the intake of any attached pump.

1301.9.5 Overflow. The storage tank shall be equipped with an overflow pipe having a diameter not less than that shown in Table 606.5.4. The overflow pipe shall be protected from insects or vermin and shall discharge in a manner consistent with storm water runoff requirements of the jurisdiction. The overflow pipe shall discharge at a sufficient distance from the tank to avoid damaging the tank foundation or the adjacent property. Drainage from overflow pipes shall be directed to prevent freezing on roof walkways. The overflow drain shall not be equipped with a shutoff valve. A cleanout shall be provided on each overflow pipe in accordance with Section 708.

1301.9.6 Access. Not less than one access opening shall be provided to allow inspection and cleaning of the tank interior. Access openings shall have an *approved* locking device or other *approved* method of securing access. Below-grade storage tanks, located outside of the building, shall be provided with a manhole either not less than 24 inches (610 mm) square or with an inside diameter not less than 24 inches (610 mm). Manholes shall extend not less than 4 inches (102 mm) above ground or shall be designed to prevent water infiltration. Finished grade shall be sloped away from the manhole to divert surface water. Manhole covers shall be secured to prevent unauthorized access. Service ports in manhole covers shall be not less than 8 inches (203 mm) in diameter and shall be not less than 4 inches (102 mm) above the finished grade level. The service port shall be secured to prevent unauthorized access.

> **Exception:** Treated-water storage tanks that are less than 800 gallons (3028 L) in volume and installed below grade shall not be required to be equipped with a manhole provided that the tank has a service port of not less than 8 inches (203 mm) in diameter.

1301.9.7 Venting. Storage tanks shall be provided with a vent sized in accordance with Chapter 9 and based on the aggregate diameter of all tank influent pipes. The reservoir vent shall not be connected to sanitary drainage system vents. Vents shall be protected from contamination by means of an *approved* cap or U-bend installed with the opening directed downward. Vent outlets shall extend not less than 4 inches (102 mm) above grade or as necessary to prevent surface water from entering the storage tank. Vent openings shall be protected against the entrance of vermin and insects in accordance with the requirements of Section 1301.7.

1301.9.8 Draining of tanks. Tanks shall be provided with a means of emptying the contents for the purpose of service or cleaning. Tanks shall be drained by using a pump or by a drain located at the lowest point in the tank. The tank drain pipe shall discharge as required for overflow pipes and shall not be smaller in size than specified in Table 606.5.7. Not less than one cleanout shall be provided on each drain pipe in accordance with Section 708.

1301.9.9 Marking and signage. Each nonpotable water storage tank shall be labeled with its rated capacity. The contents of storage tanks shall be identified with the words "CAUTION: NONPOTABLE WATER – DO NOT DRINK." Where an opening is provided that could allow the entry of personnel, the opening shall be marked with the words, "DANGER – CONFINED SPACE." Markings shall be indelibly printed on the tank or on a tag or sign constructed of corrosion-resistant waterproof material that is mounted on the tank. The letters of the words shall be not less than 0.5 inch (12.7 mm) in height and shall be of a color in contrast with the background on which they are applied.

1301.9.10 Storage tank tests. Storage tanks shall be tested in accordance with the following:

Storage tanks shall be filled with water to the overflow line prior to and during inspection. Seams and joints shall be left exposed and the tank shall remain watertight without leakage for a period of 24 hours.

1. After 24 hours, supplemental water shall be introduced for a period of 15 minutes to verify proper drainage of the overflow system and that there are no leaks.
2. The tank drain shall be observed for proper operation.
3. The makeup water system shall be observed for proper operation and successful automatic shut-off of the system at the refill threshold shall be verified.

1301.10 System abandonment. If the owner of an on-site nonpotable water reuse system or rainwater collection and conveyance system elects to cease use of, or fails to properly maintain such system, the system shall be abandoned and shall comply with the following:

1. All system piping connecting to a utility-provided water system shall be removed or disabled.
2. The distribution piping system shall be replaced with an *approved* potable water supply piping system. Where an existing potable pipe system is already in place, the fixtures shall be connected to the existing system.
3. The storage tank shall be secured from accidental access by sealing or locking tank inlets and access points, or filling with sand or equivalent.

1301.11 Trenching requirements for nonpotable water piping. Nonpotable water collection and distribution piping and reclaimed water piping shall be separated from the *building sewer* and potable water piping underground by 5 feet (1524 mm) of undisturbed or compacted earth. Nonpotable water collection and distribution piping shall not be located in, under or above cesspools, septic tanks, septic tank drainage fields or seepage pits. Buried nonpotable water piping shall comply with the requirements of Section 306.

Exceptions:

1. The required separation distance shall not apply where the bottom of the nonpotable water pipe within 5 feet (1524 mm) of the *sewer* is not less than 12 inches (305 mm) above the top of the highest point of the *sewer* and the pipe materials conform to Table 702.3.
2. The required separation distance shall not apply where the bottom of the potable water service pipe within 5 feet (1524 mm) of the nonpotable water pipe is not less than 12 inches (305 mm) above the top of the highest point of the nonpotable water pipe and the pipe materials comply with the requirements of Table 605.4.
3. Nonpotable water pipe is permitted to be located in the same trench with a *building sewer*, provided that such *sewer* is constructed of materials that comply with the requirements of Table 702.2.
4. The required separation distance shall not apply where a nonpotable water pipe crosses a *sewer* pipe, provided that the pipe is sleeved to not less than 5 feet (1524 mm) horizontally from the *sewer* pipe centerline on both sides of such crossing, with pipe materials that comply with Table 702.2.
5. The required separation distance shall not apply where a potable water service pipe crosses a nonpotable water pipe, provided that the potable water service pipe is sleeved for a distance of not less than 5 feet (1524 mm) horizontally from the centerline of the nonpotable pipe on both sides of such crossing, with pipe materials that comply with Table 702.2.
6. Irrigation piping located outside of a building and downstream of the backflow preventer is not required to meet the trenching requirements where nonpotable water is used for outdoor applications.

1301.12 Outdoor outlet access. Sillcocks, hose bibbs, wall hydrants, yard hydrants and other outdoor outlets supplied by nonpotable water shall be located in a locked vault or shall be operable only by means of a removable key.

SECTION 1302
ON-SITE NONPOTABLE WATER REUSE SYSTEMS

1302.1 General. The provisions of ASTM E2635 and Section 1302 shall govern the construction, installation, alteration and repair of on-site nonpotable water reuse systems for the collection, storage, treatment and distribution of on-site sources of nonpotable water as permitted by the jurisdiction.

1302.2 Sources. On-site nonpotable water reuse systems shall collect waste discharge from only the following sources: bathtubs, showers, lavatories, clothes washers and laundry trays. Where *approved* and as appropriate for the intended application, water from other nonpotable sources shall be collected for reuse by on-site nonpotable water reuse systems,

1302.2.1 Prohibited sources. Wastewater containing urine or fecal matter shall not be diverted to on-site nonpotable water reuse systems and shall discharge to the sanitary drainage system of the building or premises in accordance with Chapter 7. Reverse osmosis system reject water, water softener discharge water, kitchen sink wastewater, dishwasher wastewater and wastewater discharged from wet-hood scrubbers shall not be collected for reuse in an on-site nonpotable water reuse system.

1302.3 Traps. Traps serving fixtures and devices discharging wastewater to on-site nonpotable water reuse systems shall comply with Section 1002.4.

1302.4 Collection pipe. On-site nonpotable water reuse systems shall utilize drainage piping *approved* for use in plumbing drainage systems to collect and convey untreated water for reuse. Vent piping *approved* for use in plumbing venting systems shall be utilized for vents in the graywater system.

Collection and vent piping materials shall comply with Section 702.

1302.4.1 Installation. Collection piping conveying untreated water for reuse shall be installed in accordance with Section 704.

1302.4.2 Joints. Collection piping conveying untreated water for reuse shall utilize joints *approved* for use with the distribution piping and appropriate for the intended applications as specified in Section 705.

1302.4.3 Size. Collection piping conveying untreated water for reuse shall be sized in accordance with drainage sizing requirements specified in Section 710.

1302.4.4 Labeling and marking. Additional marking of collection piping conveying untreated water for reuse shall not be required beyond that required for sanitary drainage, waste and vent piping by Chapter 7.

1302.5 Filtration. Untreated water collected for reuse shall be filtered as required for the intended end use. Filters shall be provided with *access* for inspection and maintenance. Filters shall utilize a pressure gauge or other *approved* method to provide indication when a filter requires servicing or replacement. Filters shall be installed with shutoff valves immediately upstream and downstream to allow for isolation during maintenance.

1302.6 Disinfection and treatment. Where the intended application for nonpotable water collected on site for reuse requires disinfection or other treatment or both, it shall be disinfected as needed to ensure that the required water quality is delivered at the point of use. Nonpotable water collected on site containing untreated graywater shall be retained in collection reservoirs for not longer than 24 hours.

1302.6.1 Graywater used for fixture flushing. Graywater used for flushing water closets and urinals shall be disinfected and treated by an on-site water reuse treatment system complying with NSF 350.

1302.7 Storage tanks. Storage tanks utilized in on-site nonpotable water reuse systems shall comply with Sections 1301.9, 1302.7.1 and 1302.7.2.

1302.7.1 Location. Storage tanks shall be located with a minimum horizontal distance between various elements as indicated in Table 1302.7.1.

1302.7.2 Outlets. Outlets shall be located not less than 4 inches (102 mm) above the bottom of the storage tank and shall not skim water from the surface.

1302.8 Valves. Valves shall be supplied on on-site nonpotable water reuse systems in accordance with Sections 1302.8.1 and 1302.8.2.

1302.8.1 Bypass valve. One three-way diverter valve listed and labeled to NSF 50 or other *approved* device shall be installed on collection piping upstream of each storage tank, or drainfield, as applicable, to divert untreated on-site reuse sources to the sanitary *sewer* to allow servicing and inspection of the system. Bypass valves shall be installed downstream of fixture traps and vent connections. Bypass valves shall be marked to indicate the direction of flow, connection and storage tank or drainfield connection. Bypass valves shall be provided with *access* that allows for removal. Two shutoff valves shall not be installed to serve as a bypass valve.

1302.8.2 Backwater valve. One or more backwater valves shall be installed on each overflow and tank drain pipe. Backwater valves shall be in accordance with Section 714.

1302.9 Pumping and control system. Mechanical equipment including pumps, valves and filters shall be easily accessible and removable in order to perform repair, maintenance and cleaning. The minimum flow rate and flow pressure delivered by the pumping system shall be appropriate for the application and in accordance with Section 604.

1302.10 Water pressure-reducing valve or regulator. Where the water pressure supplied by the pumping system exceeds 80 psi (552 kPa) static, a pressure-reducing valve shall be installed to reduce the pressure in the nonpotable water distribution system piping to 80 psi (552 kPa) static or less. Pressure-reducing valves shall be specified and installed in accordance with Section 604.8.

1302.11 Distribution pipe. Distribution piping utilized in on-site nonpotable water reuse systems shall comply with Sections 1302.11.1 through 1302.11.3.

Exception: Irrigation piping located outside of the building and downstream of a backflow preventer.

1302.11.1 Materials, joints and connections. Distribution piping shall conform to the standards and requirements specified in Section 605.

TABLE 1302.7.1
LOCATION OF NONPOTABLE WATER REUSE STORAGE TANKS

ELEMENT	MINIMUM HORIZONTAL DISTANCE FROM STORAGE TANK (feet)
Critical root zone (CRZ) of protected trees	2
Lot line adjoining private lots	5
Seepage pits	5
Septic tanks	5
Water wells	50
Streams and lakes	50
Water service	5
Public water main	10

For SI: 1 foot = 304.8 mm.

1302.11.2 Design. On-site nonpotable water reuse distribution piping systems shall be designed and sized in accordance with Section 604 for the intended application.

1302.11.3 Marking. On-site nonpotable water distribution piping labeling and marking shall comply with Section 608.9.

1302.12 Tests and inspections. Tests and inspections shall be performed in accordance with Sections 1302.12.1 through 1302.12.6.

1302.12.1 Collection pipe and vent test. Drain, waste and vent piping used for on-site water reuse systems shall be tested in accordance with Section 312.

1302.12.2 Storage tank test. Storage tanks shall be tested in accordance with Section 1301.9.10.

1302.12.3 Water supply system test. The testing of makeup water supply piping and distribution piping shall be conducted in accordance with Section 312.5.

1302.12.4 Inspection and testing of backflow prevention assemblies. The testing of backflow preventers and backwater valves shall be conducted in accordance with Section 312.10.

1302.12.5 Inspection of vermin and insect protection. Inlets and vents to the system shall be inspected to verify that each is protected to prevent the entrance of insects and vermin into the storage tank and piping systems in accordance with Section 1301.7.

1302.12.6 Water quality test. The quality of the water for the intended application shall be verified at the point of use in accordance with the requirements of the jurisdiction.

1302.13 Operation and maintenance manuals. Operation and maintenance materials shall be supplied with nonpotable on-site water reuse systems in accordance with Sections 1302.13.1 through 1302.13.4.

1302.13.1 Manual. A detailed operations and maintenance manual shall be supplied in hardcopy form with all systems.

1302.13.2 Schematics. The manual shall include a detailed system schematic, and the locations and a list of all system components, including manufacturer and model number.

1302.13.3 Maintenance procedures. The manual shall provide a schedule and procedures for all system components requiring periodic maintenance. Consumable parts, including filters, shall be noted along with part numbers.

1302.13.4 Operations procedures. The manual shall include system startup and shutdown procedures. The manual shall include detailed operating procedures for the system.

SECTION 1303
NONPOTABLE RAINWATER COLLECTION AND DISTRIBUTION SYSTEMS

1303.1 General. The provisions of Section 1303 shall govern the construction, installation, alteration and repair of rainwater collection and conveyance systems for the collection, storage, treatment and distribution of rainwater for nonpotable applications, as permitted by the jurisdiction.

1303.1.1 Fire protection systems. The storage, treatment and distribution of nonpotable water to be used for fire protection systems shall be in accordance with the *International Fire Code*.

1303.2 Collection surface. Rainwater shall be collected only from above-ground impervious roofing surfaces constructed from *approved* materials and where *approved*, vehicular parking or pedestrian walking surfaces.

1303.3 Debris excluders. Downspouts and leaders shall be connected to a debris excluder or equivalent device that is designed to remove leaves, sticks, pine needles and similar debris to prevent such from entering the storage tank.

1303.4 First-flush diverter. First-flush diverters shall operate automatically and shall not rely on manually operated valves or devices. Diverted rainwater shall not be drained to the roof surface, and shall be discharged in a manner consistent with the storm water runoff requirements of the jurisdiction. First-flush diverters shall be provided with *access* for maintenance and service.

1303.5 Roof gutters and downspouts. Gutters and downspouts shall be constructed of materials that are compatible with the collection surface and the rainwater quality for the desired end use. Joints shall be watertight.

1303.5.1 Slope. Roof gutters, leaders and rainwater collection piping shall slope continuously toward collection inlets. Gutters and downspouts shall have a slope of not less than $1/_8$ inch per foot (10.4 mm/m) along their entire length, and shall not permit the collection or pooling of water at any point.

> **Exception:** Siphonic drainage systems installed in accordance with the manufacturer's instructions shall not be required to have a slope.

1303.5.2 Size. Gutters and downspouts shall be installed and sized in accordance with Section 1106.6 and local rainfall rates.

1303.5.3 Cleanouts. Cleanouts shall be provided in the water conveyance system to allow *access* to all filters, flushes, pipes and downspouts.

1303.6 Drainage. Water drained from the roof washer or debris excluder shall not be drained to the sanitary sewer. Such water shall be diverted from the storage tank and discharge in a location that will not cause erosion or damage to property in accordance with the *International Building Code*. Roof washers and debris excluders shall be provided with an automatic means of self-draining between rain events, and shall not drain onto roof surfaces.

1303.7 Collection pipe. Rainwater collection and conveyance systems shall utilize drainage piping *approved* for use within plumbing drainage systems to collect and convey captured rainwater. Vent piping *approved* for use within plumbing venting systems shall be utilized for vents within the rainwater system. Collection and vent piping materials shall comply with Section 702.

1303.7.1 Installation. Collection piping conveying captured rainwater shall be installed in accordance with Section 704.

1303.7.2 Joints. Collection piping conveying captured rainwater shall utilize joints *approved* for use with the distribution piping and appropriate for the intended applications as specified in Section 705.

1303.7.3 Size. Collection piping conveying captured rainwater shall be sized in accordance with drainage sizing requirements specified in Section 710.

1303.7.4 Marking. Additional marking of collection piping conveying captured rainwater for reuse shall not be required beyond that required for sanitary drainage, waste and vent piping by Chapter 7.

1303.8 Filtration. Collected rainwater shall be filtered as required for the intended end use. Filters shall be provided with *access* for inspection and maintenance. Filters shall utilize a pressure gauge or other *approved* method to provide indication when a filter requires servicing or replacement. Filters shall be installed with shutoff valves installed immediately upstream and downstream to allow for isolation during maintenance.

1303.9 Disinfection. Where the intended application for rainwater requires disinfection or other treatment or both, it shall be disinfected as needed to ensure that the required water quality is delivered at the point of use. Where chlorine is used for disinfection or treatment, water shall be tested for residual chlorine in accordance with ASTM D1253. The levels of residual chlorine shall not exceed that allowed for the intended use in accordance with the requirements of the jurisdiction.

1303.10 Storage tanks. Storage tanks utilized in nonpotable rainwater collection and conveyance systems shall comply with Sections 1301.9 and 1303.10.1 through 1303.10.3.

1303.10.1 Location. Storage tanks shall be located with a minimum horizontal distance between various elements as indicated in Table 1303.10.1.

1303.10.2 Inlets. Storage tank inlets shall be designed to introduce collected rainwater into the tank with minimum turbulence, and shall be located and designed to avoid agitating the contents of the storage tank.

1303.10.3 Outlets. Outlets shall be located not less than 4 inches (102 mm) above the bottom of the storage tank and shall not skim water from the surface.

1303.11 Valves. Valves shall be supplied on rainwater collection and conveyance systems in accordance with Section 1303.11.1.

1303.11.1 Backwater valve. Backwater valves shall be installed on each overflow and tank drain pipe. Backwater valves shall be in accordance with Section 714.

1303.12 Pumping and control system. Mechanical equipment including pumps, valves and filters shall be provided with *access* that allows for removal in order to perform repair, maintenance and cleaning. The minimum flow rate and flow pressure delivered by the pumping system shall be appropriate for the application and in accordance with Section 604.

1303.13 Water pressure-reducing valve or regulator. Where the water pressure supplied by the pumping system exceeds 80 psi (552 kPa) static, a pressure-reducing valve shall be installed to reduce the pressure in the rainwater distribution system piping to 80 psi (552 kPa) static or less. Pressure-reducing valves shall be specified and installed in accordance with Section 604.8.

1303.14 Distribution pipe. Distribution piping utilized in rainwater collection and conveyance systems shall comply with Sections 1303.14.1 through 1303.14.3.

Exception: Irrigation piping located outside of the building and downstream of a backflow preventer.

1303.14.1 Materials, joints and connections. Distribution piping shall conform to the standards and requirements specified in Section 605 for nonpotable water.

1303.14.2 Design. Distribution piping systems shall be designed and sized in accordance with Section 604 for the intended application.

1303.14.3 Marking. Nonpotable rainwater distribution piping labeling and marking shall comply with Section 608.9.

1303.15 Tests and inspections. Tests and inspections shall be performed in accordance with Sections 1303.15.1 through 1303.15.9.

1303.15.1 Roof gutter inspection and test. Roof gutters shall be inspected to verify that the installation and slope is in accordance with Section 1303.5.1. Gutters shall be tested by pouring not less than 1 gallon (3.8 l) of water into the end of the gutter opposite the collection point. The gutter being tested shall not leak and shall not retain standing water.

1303.15.2 First-flush diverter test. First-flush diverters shall be tested by introducing water into the collection system upstream of the diverter. Proper diversion of the first amount of water shall be in accordance with the requirements of Section 1303.4.

TABLE 1303.10.1
LOCATION OF RAINWATER STORAGE TANKS

ELEMENT	MINIMUM HORIZONTAL DISTANCE FROM STORAGE TANK (feet)
Critical root zone (CRZ) of protected trees	2
Lot line adjoining private lots	5
Seepage pits	5
Septic tanks	5

For SI: 1 foot = 304.8 mm.

1303.15.3 Collection pipe and vent test. Drain, waste and vent piping used for rainwater collection and conveyance systems shall be tested in accordance with Section 312.

1303.15.4 Storage tank test. Storage tanks shall be tested in accordance with Section 1301.9.10.

1303.15.5 Water supply system test. The testing of makeup water supply piping and distribution piping shall be conducted in accordance with Section 312.5.

1303.15.6 Inspection and testing of backflow prevention assemblies. The testing of backflow preventers and backwater valves shall be conducted in accordance with Section 312.10.

1303.15.7 Inspection of vermin and insect protection. Inlets and vents to the system shall be inspected to verify that each is protected to prevent the entrance of insects and vermin into the storage tank and piping systems in accordance with Section 1301.7.

1303.15.8 Water quality test. The quality of the water for the intended application shall be verified at the point of use in accordance with the requirements of the jurisdiction.

1303.15.9 Collected raw rainwater quality. ASTM E2727 shall be used to determine what, if any, site conditions impact the quality of collected raw rainwater and whether those site conditions require treatment of the raw water for the intended end use or make the water unsuitable for specific end uses.

1303.16 Operation and maintenance manuals. Operation and maintenance manuals shall be supplied with rainwater collection and conveyance systems in accordance with Sections 1303.16.1 through 1303.16.4.

1303.16.1 Manual. A detailed operations and maintenance manual shall be supplied in hardcopy form with all systems.

1303.16.2 Schematics. The manual shall include a detailed system schematic, and locations and a list of all system components, including manufacturer and model number.

1303.16.3 Maintenance procedures. The manual shall provide a maintenance schedule and procedures for all system components requiring periodic maintenance. Consumable parts, including filters, shall be noted along with part numbers.

1303.16.4 Operations procedures. The manual shall include system startup and shutdown procedures, as well as detailed operating procedures.

SECTION 1304
RECLAIMED WATER SYSTEMS

1304.1 General. The provisions of this section shall govern the construction, installation, alteration and repair of systems supplying nonpotable reclaimed water.

1304.2 Water pressure-reducing valve or regulator. Where the reclaimed water pressure supplied to the building exceeds 80 psi (552 kPa) static, a pressure-reducing valve shall be installed to reduce the pressure in the reclaimed water distribution system piping to 80 psi (552 kPa) static or less. Pressure-reducing valves shall be specified and installed in accordance with Section 604.8.

1304.3 Reclaimed water systems. The design of the reclaimed water systems shall conform to *accepted engineering practice*.

1304.3.1 Distribution pipe. Distribution piping shall comply with Sections 1304.3.1.1 through 1304.3.1.3.

Exception: Irrigation piping located outside of the building and downstream of a backflow preventer.

1304.3.1.1 Materials, joints and connections. Distribution piping conveying reclaimed water shall conform to standards and requirements specified in Section 605 for nonpotable water.

1304.3.1.2 Design. Distribution piping systems shall be designed and sized in accordance with Section 604 for the intended application.

1304.3.1.3 Labeling and marking. Nonpotable distribution piping labeling and marking shall comply with Section 608.9.

1304.4 Tests and inspections. Tests and inspections shall be performed in accordance with Sections 1304.4.1 and 1304.4.2.

1304.4.1 Water supply system test. The testing of makeup water supply piping and reclaimed water distribution piping shall be conducted in accordance with Section 312.5.

1304.4.2 Inspection and testing of backflow prevention assemblies. The testing of backflow preventers shall be conducted in accordance with Section 312.10.

CHAPTER 14

SUBSURFACE LANDSCAPE IRRIGATION SYSTEMS

User note:

About this chapter: Chapter 14 provides regulations for disposing of nonpotable water to underground landscape irrigation piping. Testing procedures are provided to assess the capability of the soil to accept the volume of flow. This chapter covers the regulations, types and material standards for the piping of these systems.

SECTION 1401
GENERAL

1401.1 Scope. The provisions of this chapter shall govern the materials, design, construction and installation of subsurface landscape irrigation systems connected to nonpotable water from on-site water reuse systems.

1401.2 Materials. Above-ground drain, waste and vent piping for subsurface landscape irrigation systems shall conform to one of the standards listed in Table 702.1. Subsurface landscape irrigation, underground building drainage and vent pipe shall conform to one of the standards listed in Table 702.2.

1401.3 Tests. Drain, waste and vent piping for subsurface landscape irrigation systems shall be tested in accordance with Section 312.

1401.4 Inspections. Subsurface landscape irrigation systems shall be inspected in accordance with Section 107.

1401.5 Disinfection. Disinfection shall not be required for on-site nonpotable water reuse for subsurface landscape irrigation systems.

1401.6 Coloring. On-site nonpotable water reuse for subsurface landscape irrigation systems shall not be required to be dyed.

SECTION 1402
SYSTEM DESIGN AND SIZING

1402.1 Sizing. The system shall be sized in accordance with the sum of the output of all water sources connected to the subsurface irrigation system. Where graywater collection piping is connected to subsurface landscape irrigation systems, graywater output shall be calculated according to the gallons-per-day-per-occupant number based on the type of fixtures connected. The graywater discharge shall be calculated by the following equation:

$C = A \times B$ **(Equation 14-1)**

where:

A = Number of occupants:

Residential—Number of occupants shall be determined by the actual number of occupants, but not less than two occupants for one bedroom and one occupant for each additional bedroom.

Commercial—Number of occupants shall be determined by the *International Building Code*.

B = Estimated flow demands for each occupant:

Residential—25 gallons per day (94.6 lpd) per occupant for showers, bathtubs and lavatories and 15 gallons per day (56.7 lpd) per occupant for clothes washers or laundry trays.

Commercial—Based on type of fixture or water use records minus the discharge of fixtures other than those discharging graywater.

C = Estimated graywater discharge based on the total number of occupants.

1402.2 Percolation tests. The permeability of the soil in the proposed absorption system shall be determined by percolation tests or permeability evaluation.

1402.2.1 Percolation tests and procedures. Not fewer than three percolation tests in each system area shall be conducted. The holes shall be spaced uniformly in relation to the bottom depth of the proposed absorption system. More percolation tests shall be made where necessary, depending on system design.

1402.2.1.1 Percolation test hole. The test hole shall be dug or bored. The test hole shall have vertical sides and a horizontal dimension of 4 inches to 8 inches (102 mm to 203 mm). The bottom and sides of the hole shall be scratched with a sharp-pointed instrument to expose the natural soil. Loose material shall be removed from the hole and the bottom shall be covered with 2 inches (51 mm) of gravel or coarse sand.

1402.2.1.2 Test procedure, sandy soils. The hole shall be filled with clear water to not less than 12 inches (305 mm) above the bottom of the hole for tests in sandy soils. The time for this amount of water to seep away shall be determined, and this procedure shall be repeated if the water from the second filling of the hole seeps away in 10 minutes or less. The test shall proceed as follows: Water shall be added to a point not more than 6 inches (152 mm) above the gravel or coarse sand. Thereupon, from a fixed reference point, water levels shall be measured at 10-minute intervals for a period of 1 hour. Where 6 inches (152 mm) of water seeps away in less than 10 minutes, a shorter interval between measurements shall be used, but in no case shall the water depth exceed 6 inches (152 mm). Where

SUBSURFACE LANDSCAPE IRRIGATION SYSTEMS

6 inches (152 mm) of water seeps away in less than 2 minutes, the test shall be stopped and a rate of less than 3 minutes per inch (7.2 s/mm) shall be reported. The final water level drop shall be used to calculate the percolation rate. Soils not meeting the requirements of this section shall be tested in accordance with Section 1402.2.1.3.

1402.2.1.3 Test procedure, other soils. The hole shall be filled with clear water, and a minimum water depth of 12 inches (305 mm) shall be maintained above the bottom of the hole for a 4-hour period by refilling whenever necessary or by use of an automatic siphon. Water remaining in the hole after 4 hours shall not be removed. Thereafter, the soil shall be allowed to swell not less than 16 hours or more than 30 hours. Immediately after the soil swelling period, the measurements for determining the percolation rate shall be made as follows: any soil sloughed into the hole shall be removed and the water level shall be adjusted to 6 inches (152 mm) above the gravel or coarse sand. Thereupon, from a fixed reference point, the water level shall be measured at 30-minute intervals for a period of 4 hours, unless two successive water level drops do not vary by more than $1/16$ inch (1.59 mm). Not fewer than three water level drops shall be observed and recorded. The hole shall be filled with clear water to a point not more than 6 inches (152 mm) above the gravel or coarse sand whenever it becomes nearly empty. Adjustments of the water level shall not be made during the three measurement periods except to the limits of the last measured water level drop. Where the first 6 inches (152 mm) of water seeps away in less than 30 minutes, the time interval between measurements shall be 10 minutes and the test run for 1 hour. The water depth shall not exceed 5 inches (127 mm) at any time during the measurement period. The drop that occurs during the final measurement period shall be used in calculating the percolation rate.

1402.2.1.4 Mechanical test equipment. Mechanical percolation test equipment shall be of an *approved* type.

1402.2.2 Permeability evaluation. Soil shall be evaluated for estimated percolation based on structure and texture in accordance with accepted soil evaluation practices. Borings shall be made in accordance with Section 1402.2.1.1 for evaluating the soil.

1402.3 Subsurface landscape irrigation site location. The surface grade of all soil absorption systems shall be located at a point lower than the surface grade of any water well or reservoir on the same or adjoining lot. Where this is not possible, the site shall be located so surface water drainage from the site is not directed toward a well or reservoir. The soil absorption system shall be located with a minimum horizontal distance between various elements as indicated in Table 1402.3. Private sewage disposal systems in compacted areas, such as parking lots and driveways, are prohibited. Surface water shall be diverted away from any soil absorption site on the same or neighboring lots.

TABLE 1402.3
LOCATION OF SUBSURFACE IRRIGATION SYSTEM

ELEMENT	MINIMUM HORIZONTAL DISTANCE	
	Storage tank (feet)	Irrigation disposal field (feet)
Buildings	5	2
Lot line adjoining private property	5	5
Water wells	50	100
Streams and lakes	50	50
Seepage pits	5	5
Septic tanks	0	5
Water service	5	5
Public water main	10	10

For SI: 1 foot = 304.8 mm.

TABLE 1403.1.1
DESIGN LOADING RATE

PERCOLATION RATE (minutes per inch)	DESIGN LOADING FACTOR (gallons per square foot per day)
0 to less than 10	1.2
10 to less than 30	0.8
30 to less than 45	0.72
45 to 60	0.4

For SI: 1 minute per inch = min/25.4 mm, 1 gallon per square foot = 40.7 L/m^2.

SECTION 1403
INSTALLATION

1403.1 Installation. Absorption systems shall be installed in accordance with Sections 1403.1.1 through 1403.1.5 to provide landscape irrigation without surfacing of water.

1403.1.1 Absorption area. The total absorption area required shall be computed from the estimated daily graywater discharge and the design-loading rate based on the percolation rate for the site. The required absorption area equals the estimated graywater discharge divided by the design-loading rate from Table 1403.1.1.

1403.1.2 Seepage trench excavations. Seepage trench excavations shall be not less than 1 foot (304 mm) in width and not greater than 5 feet (1524 mm) in width. Trench excavations shall be spaced not less than 2 feet (610 mm) apart. The soil absorption area of a seepage trench shall be computed by using the bottom of the trench area (width) multiplied by the length of pipe. Individual seepage trenches shall be not greater than 100 feet (30 480 mm) in *developed length*.

1403.1.3 Seepage bed excavations. Seepage bed excavations shall be not less than 5 feet (1524 mm) in width and have more than one distribution pipe. The absorption area of a seepage bed shall be computed by using the bottom of the trench area. Distribution piping in a seepage bed shall be uniformly spaced not greater than 5 feet (1524 mm) and not less than 3 feet (914 mm) apart, and greater than 3 feet (914 mm) and not less than 1 foot (305 mm) from the sidewall or headwall.

1403.1.4 Excavation and construction. The bottom of a trench or bed excavation shall be level. Seepage trenches or beds shall not be excavated where the soil is so wet that such material rolled between the hands forms a soil wire. Smeared or compacted soil surfaces in the sidewalls or bottom of seepage trench or bed excavations shall be scarified to the depth of smearing or compaction and the loose material removed. Where rain falls on an open excavation, the soil shall be left until sufficiently dry so a soil wire will not form when soil from the excavation bottom is rolled between the hands. The bottom area shall then be scarified and loose material removed.

1403.1.5 Aggregate and backfill. Not less than 6 inches in depth of aggregate, ranging in size from $^1/_2$ to $2^1/_2$ inches (12.7 mm to 64 mm), shall be laid into the trench below the distribution piping elevation. The aggregate shall be evenly distributed not less than 2 inches (51 mm) in depth over the top of the distribution pipe. The aggregate shall be covered with *approved* synthetic materials or 9 inches (229 mm) of uncompacted marsh hay or straw. Building paper shall not be used to cover the aggregate. Not less than 9 inches (229 mm) of soil backfill shall be provided above the covering.

1403.2 Distribution piping. Distribution piping shall be not less than 3 inches (76 mm) in diameter. Materials shall comply with Table 1403.2. The top of the distribution pipe shall be not less than 8 inches (203 mm) below the original surface. The slope of the distribution pipes shall be not less than 2 inches (51 mm) and not greater than 4 inches (102 mm) per 100 feet (30 480 mm).

1403.2.1 Joints. Joints in distribution pipe shall be made in accordance with Section 705 of this code.

TABLE 1403.2
DISTRIBUTION PIPE

MATERIAL	STANDARD
Polyethylene (PE) plastic pipe	ASTM F405
Polyvinyl chloride (PVC) plastic pipe	ASTM D2729
Polyvinyl chloride (PVC) plastic pipe with a 3.5-inch O.D. and solid cellular core or composite wall	ASTM F1488

For SI: 1 inch = 25.4 mm.

CHAPTER 15
REFERENCED STANDARDS

User note:

About this chapter: This code contains numerous references to standards that are used to provide requirements for materials and methods of construction. Chapter 15 contains a comprehensive list of all standards that are referenced in this code. These standards, in essence, are part of this code to the extent of the reference to the standard.

This chapter lists the standards that are referenced in various sections of this document. The standards are listed herein by the promulgating agency of the standard, the standard identification, the effective date and title, and the section or sections of this document that reference the standard. The application of the referenced standards shall be as specified in Section 102.8.

ANSI

American National Standards Institute
25 West 43rd Street, 4th Floor
New York, NY 10036

A118.10—99: Specifications for Load Bearing, Bonded, Waterproof Membranes for Thin Set Ceramic Tile and Dimension Stone Installation
    421.5.2.5, 421.5.2.6

Z21.22—99 (R2003): Relief Valves for Hot Water Supply Systems with Addenda Z21.22a—2000 (R2003) and Z21.22b—2001 (R2003)
    504.2, 504.4, 504.4.1

ASHRAE

ASHRAE
1791 Tullie Circle NE
Atlanta, GA 30329

ASHRAE 18—2008 (RA13): Method of Testing for Rating Drinking-Water Coolers with Self-contained Mechanical Refrigeration (ANSI/ASHRAE Approved)
    410.1

ASME

American Society of Mechanical Engineers
Two Park Avenue
New York, NY 10016-5990

A112.1.2—2012: Air Gaps in Plumbing Systems (For Plumbing Fixtures and Water Connection Receptors)
    406.1, 409.2, Table 608.1, 608.14.1

A112.1.3—2000 (R2015): Air Gap Fittings for Use with Plumbing Fixtures, Appliances and Appurtenances
    406.1, 409.2, Table 608.1, 608.14.1, 1102.6

A112.3.1—2007 (R2012): Stainless Steel Drainage Systems for Sanitary, DWV, Storm and Vacuum Applications Above and Below Ground
    413.1, Table 702.1, Table 702.2, Table 702.3, Table 702.4, Table 1102.4, Table 1102.5, 1102.6, Table 1102.7

ASME A112.3.4—2013/CSA B45.9—2013: Macerating Toilet Systems and Related Components
    405.5, 712.4.1

A112.4.1—2009: Water Heater Relief Valve Drain Tubes
    504.6

A112.4.2—2015/CSA B45.16—15: Water Closet Personal Hygiene Devices
    412.9

A112.4.3—1999 (R2010): Plastic Fittings for Connecting Water Closets to the Sanitary Drainage System
    405.4

A112.4.14—2004 (R2016): Manually Operated, Quarter-turn Shutoff Valves for Use in Plumbing Systems
    Table 605.7

A112.6.2—2000 (R2016): Framing-affixed Supports for Off-the-floor Water Closets with Concealed Tanks
    405.4.3

REFERENCED STANDARDS

ASME—continued

A112.6.3—2001 (R2016): Floor and Trench Drains
    413.1

A112.6.4—2003 (R2012): Roof, Deck, and Balcony Drains
    1102.6

A112.6.7—2010 (R2015): Sanitary Floor Sinks
    414.1

A112.6.9—2005 (R2015): Siphonic Roof Drains
    1107.1

A112.14.1—2003 (R2012): Backwater Valves
    714.2

A112.14.3—2016: Grease Interceptors
    1003.3.5

A112.14.4—2001 (R2012): Grease Removal Devices
    1003.3.5

A112.14.6—2010 (R2015): FOG (Fats, Oils and Greases) Disposal Systems
    1003.3, 1003.3.5, 1003.3.7

A112.18.1—2017/CSA B125.1—2017: Plumbing Supply Fittings
    412.1, 412.2, 412.3, 412.4, 412.6, 412.8, 605.7, 607.4, 608.2

A112.18.2—2015/CSA B125.2—15: Plumbing Waste Fittings
    412.1.2

A112.18.3—2002 (R2012): Performance Requirements for Backflow Protection Devices and Systems in Plumbing Fixture Fittings
    412.2, 412.6

A112.18.6—2017/CSA B125.6—17: Flexible Water Connectors
    605.6

A112.18.9—2011: Protectors/Insulators for Exposed Waste and Supplies on Accessible Fixtures
    404.3

A112.19.1—2013/CSA B45.2—2013: Enameled Cast Iron and Enameled Steel Plumbing Fixtures
    407.1, 410.1, 418.1, 419.1, 422.1

A112.19.2—2013/CSA B45.1—13: Ceramic Plumbing Fixtures
    401.2, 405.10, 407.1, 408.1, 410.1, 418.1, 419.1, 421.1, 422.1, 424.1, 425.1

A112.19.3—2008/CSA B45.4—08 (R2013): Stainless Steel Plumbing Fixtures
    405.9, 407.1, 418.1, 419.1, 425.1

A112.19.5—2017/CSA B45.15—2017: Flush Valves and Spuds for Water-closets, Urinals, and Tanks
    415.4

A112.19.7M—2017/CSA B45.10—17: Hydromassage Bathtub Systems
    426.1, 426.4

A112.19.12—2014: Wall Mounted and Pedestal Mounted, Adjustable, Elevating, Tilting and Pivoting Lavatory, Sink and Shampoo Bowl Carrier Systems and Drain Waste Systems
    418.3, 419.4

A112.19.14-2013: Six-liter Water Closets Equipped with a Dual Flushing Device
    425.1

A112.19.15—2012: Bathtub/Whirlpool Bathtubs with Pressure Sealed Doors
    407.4, 426.6

A112.19.19—2006 (R2011): Vitreous China Nonwater Urinals
    424.1

A112.21.3-1985(R2007): Hydrants for Utility and Maintenance Use
    Table 608.1, 608.14.6

A112.36.2M—1991(R2012): Cleanouts
    708.1.10.2

ASSE 1002—2015/ASME A112.1002—2015/CSA B125.12—15: Anti-Siphon Fill Valves
    415.3.1, Table 608.1

ASME—continued

ASSE 1016—2017/ASME A112.1016—2017/CSA B125.16—2017: Performance Requirements for Individual Thermostatic, Pressure Balancing and Combination Control Valves for Individual Fixture Fittings
    412.3, 412.4, 607.4

ASSE 1037—2015/ASME A112.1037—2015/CSA B125.37—: Pressurized Flushing Devices for Plumbing Fixtures
    415.2

ASSE 1070—2015/ASME A112.1070—2015/CSA B125.1070—15: Water Temperature Limiting Devices
    408.3, 412.5, 412.10, 419.5, 423.3, 607.1.2

B1.20.1—2013: Pipe Threads, General Purpose (inch)
    605.10.3, 605.12.4, 605.14.3, 605.17.1, 605.22.4, 705.2.3, 705.5.4, 705.8.1, 705.10.3, 705.19.1

B16.3—2016: Malleable Iron Threaded Fittings Classes 150 and 300
    Table 1102.7

B16.4—2016: Gray Iron Threaded Fittings Classes 125 and 250
    Table 605.5, Table 702.4, Table 1102.7

B16.9—2012: Factory-made Wrought Steel Buttwelding Fittings
    Table 605.5, Table 702.4, Table 1102.7

B16.11—2016: Forged Fittings, Socket-welding and Threaded
    Table 605.5, Table 702.4, Table 1102.7

B16.12—2009 (R2014): Cast-iron Threaded Drainage Fittings
    Table 702.4, Table 1102.7

B16.15—2013: Cast Alloy Threaded Fittings: Class 125 and 250
    Table 605.5, Table 702.4, Table 1102.7

B16.18—2012: Cast Copper Alloy Solder Joint Pressure Fittings
    Table 605.5, Table 702.4, Table 1102.7

B16.22—2013: Wrought Copper and Copper Alloy Solder Joint Pressure Fittings
    Table 605.5, Table 702.4, Table 1102.7

B16.23—2016: Cast Copper Alloy Solder Joint Drainage Fittings DWV
    Table 702.4, Table 1102.7

B16.26—2016: Cast Copper Alloy Fittings for Flared Copper Tubes
    Table 605.5, Table 702.4, Table 1102.7

B16.28—1994: Wrought Steel Buttwelding Short Radius Elbows and Returns
    Table 605.5, Table 702.4, Table 1102.7

B16.29—2012: Wrought Copper and Wrought Copper Alloy Solder Joint Drainage Fittings (DWV)
    Table 702.4, Table 1102.7

B16.34—2015: Valves Flanged, Threaded and Welding End
    Table 605.7

B16.51—2013: Copper and Copper Alloy Press-connect Pressure Fittings
    Table 605.5

ASPE

American Society of Plumbing Engineers
6400 Shafer Ct., Suite 350
Rosemont, IL 60018-4914

45—2013: Siphonic Roof Drainage Systems
    1107.1

ASSE

ASSE International
18927 Hickory Creek Drive, Suite 220
Mokena, IL 60448

1001—2016: Performance Requirements for Atmospheric Type Vacuum Breakers
    415.2, Table 608.1, 608.14.6, 608.17.4.1

ASSE 1002—2015/ASME A112.1002—2015/CSA B125.12—15: Antisiphon Fill Valves
    415.3.1, Table 608.1

REFERENCED STANDARDS

ASSE—continued

1003—09: Performance Requirements for Water Pressure Reducing Valves
604.8

1004—2016: Performance Requirements for Commercial Dishwashing Machines
409.1

1005—99: Performance Requirements for Water Heater Drain Valves
501.3

1008—06: Performance Requirements for Plumbing Aspects of Food Waste Disposer Units
413.1

1010—04: Performance Requirements for Water Hammer Arresters
604.9

1011—2016: Performance Requirements for Hose Connection Vacuum Breakers
Table 608.1, 608.14.6

1012—09: Performance Requirements for Backflow Preventers with Intermediate Atmospheric Vent
Table 608.1, 608.14.3, 608.17.2, 608.17.10

1013—2017: Performance Requirements for Reduced Pressure Principle Backflow Preventers and Reduced Pressure Principle Fire Protection Backflow Preventers
Table 608.1, 608.14.2, 608.17.2

1015—2017: Performance Requirements for Double Check Backflow Prevention Assemblies and Double Check Fire Protection Backflow Prevention Assemblies
Table 608.1, 608.14.7

ASSE 1016—2017/ASME A112.1016—2017/CSA B125.16—2017: Performance Requirements for Individual Thermostatic, Pressure Balancing and Combination Control Valves for Individual Fixture Fittings
412.3, 412.4, 607.4

1017—2010: Performance Requirements for Temperature Actuated Mixing Valves for Hot Water Distribution Systems
501.2, 613.1

1018—2017: Performance Requirements for Trap Seal Primer Valves; Potable Water Supplied
1002.4.1.2

1019—2016: Performance Requirements for Vacuum Breaker Wall Hydrants, Freeze Resistant, Automatic Draining Type
Table 608.1, 608.14.6

1020—04: Performance Requirements for Pressure Vacuum Breaker Assembly
Table 608.1, 608.14.5

1022—2016: Performance Requirements for Backflow Preventer for Beverage Dispensing Equipment
Table 608.1, 608.17.1.1, 608.17.1.2

1024—2016: Performance Requirements for Dual Check Valve Type Backflow Preventers, Anti-siphon-type, Residential Applications
605.3.1, Table 608.1, 608.14.9, 608.17.1.2

1035—08: Performance Requirements for Laboratory Faucet Backflow Preventers
Table 608.1, 608.14.6

ASSE 1037—2015/ASME A112.1037—2015/CSA B125.37—15: Pressurized Flushing Devices for Plumbing Fixtures
415.2

1044—2010: Performance Requirements for Trap Seal Primer Devices Drainage Types and Electronic Design Types
1002.4.1.3

1047—2017: Performance Requirements for Reduced Pressure Detector Fire Protection Backflow Prevention Assemblies
Table 608.1, 608.14.2

1048—2017: Performance Requirements for Double Check Detector Fire Protection Backflow Prevention Assemblies
Table 608.1, 608.14.7

1049—2009: Performance Requirements for Individual and Branch Type Air Admittance Valves for Chemical Waste Systems
901.3, 918.8

1050—2009: Performance Requirements for Stack Air Admittance Valves for Sanitary Drainage Systems
918.1

1051—2009: Performance Requirements for Individual and Branch Type Air Admittance Valves for Sanitary Drainage Systems-fixture and Branch Devices
918.1

ASSE—continued

1052—2016: Performance Requirements for Hose Connection Backflow Preventers
Table 608.1, 608.14.6

1055—2016: Performance Requirements for Chemical Dispensing Systems
608.14.8

1056—2013: Performance Requirements for Spill Resistant Vacuum Breaker
Table 608.1, 608.14.5

1060—2016: Performance Requirements for Outdoor Enclosures for Fluid Conveying Components
608.15.1

1061—2015: Performance Requirements for Push Fit Fittings
Table 605.5, 604.13.7, 605.14.4, 605.16.3

1062—2016: Performance Requirements for Temperature Actuated, Flow Reduction (TAFR) Valves to Individual Supply Fittings
424.7

1066—2016: Performance Requirements for Individual Pressure Balancing In-line Valves for Individual Fixture Fittings
604.11

1069—05: Performance Requirements for Automatic Temperature Control Mixing Valves
412.4

ASSE 1070—2015/ASME A112.1070—2015/CSA B125.70—15: Water-temperature Limiting Devices
408.3, 419.5, 423.3, 424.5, 607.1.2

1071—2012: Performance Requirements for Temperature Actuated Mixing Valves for Plumbed Emergency Equipment
411.3, 412.5, 412.10, 419.5, 423.3, 607.1.2

1072—07: Performance Requirements for Barrier Type Floor Drain Tap Seal Protection Devices
1002.4.14

1079—2005: Performance Requirements for Dielectric Pipe Unions
605.23.1, 605.23.3

5013—2015: Performance Requirements for Testing Reduced Pressure Principle Backflow Prevention Assembly (RPA) and Reduced Pressure Fire Protection Backflow Preventers (RFP)
312.10.2

5015—2015: Performance Requirements for Testing Double Check Valve Backflow Prevention Assemblies (DC) and Double Check Fire Protection Backflow Prevention Assemblies (DCF)
312.10.2

5020—2015: Performance Requirements for Testing Pressure Vacuum Breaker Assemblies (PVBA)
312.10.2

5047—2015: Performance Requirements for Testing Reduced Pressure Detector Fire Protection Backflow Prevention Assemblies (RPDA)
312.10.2

5048—2015: Performance Requirements for Testing Double Check Valve Detector Assembly (DCDA)
312.10.2

5052—98: Performance Requirements for Testing Hose Connection Backflow Preventers
312.10.2

5056—2015: Performance Requirements for Testing Spill Resistant Vacuum Breaker (SRVB)
312.10.2

ASTM

ASTM International
100 Barr Harbor Drive, P.O. Box C700
West Conshohocken, PA 19428-2959

A53/A53M—12: Specification for Pipe, Steel, Black and Hot-dipped, Zinc-coated Welded and Seamless
Table 605.3, Table 605.4, Table 702.1

A74—15: Specification for Cast-iron Soil Pipe and Fittings
Table 702.1, Table 702.2, Table 702.3, Table 702.4, 708.1.6, 708.7, Table 1102.4, Table 1102.5, Table 1102.7

A312/A312M—15a: Specification for Seamless, Welded, and Heavily Cold Worked Austenitic Stainless Steel Pipes
Table 605.3, Table 605.4, Table 605.5, 605.22.2

REFERENCED STANDARDS

ASTM—continued

A733—15: Specification for Welded and Seamless Carbon Steel and Austenitic Stainless Steel Pipe Nipples
Table 605.8

A778/A778M—15: Specification for Welded Unannealed Austenitic Stainless Steel Tubular Products
Table 605.3, Table 605.4, Table 605.5

A888—15: Specification for Hubless Cast-iron Soil Pipe and Fittings for Sanitary and Storm Drain, Waste, and Vent Piping Application
Table 702.1, Table 702.2, Table 702.3, Table 702.4, 708.7, Table 1102.4, Table 1102.5, Table 1102.7

B32—08(2014): Specification for Solder Metal
605.12.3, 605.13.6, 705.5.3, 705.6.1, 705.17.3

B42—15a: Specification for Seamless Copper Pipe, Standard Sizes
Table 605.3, Table 605.4, Table 702.1

B43—15: Specification for Seamless Red Brass Pipe, Standard Sizes
Table 605.4, Table 702.1

B75/B75M—11: Specification for Seamless Copper Tube
Table 605.3, Table 605.4, Table 702.1, Table 702.2, Table 702.3, Table 1102.4

B88—14: Specification for Seamless Copper Water Tube
Table 605.3, Table 605.4, Table 702.1, Table 702.2, Table 702.3, Table 1102.4

B152/B152M—13: Specification for Copper Sheet, Strip Plate and Rolled Bar
402.3, 415.3.3, 422.5.2.4, 902.2

B251—10: Specification for General Requirements for Wrought Seamless Copper and Copper-alloy Tube
Table 605.3, Table 605.4, Table 702.1, Table 702.2, Table 702.3, Table 1102.4

B302—12: Specification for Threadless Copper Pipe, Standard Sizes
Table 605.3, Table 605.4, Table 702.1

B306—13: Specification for Copper Drainage Tube (DWV)
Table 702.1, Table 702.2, Table 1102.4

B447—12a: Specification for Welded Copper Tube
Table 605.3, Table 605.4

B687—99 (2011): Specification for Brass, Copper and Chromium-plated Pipe Nipples
Table 605.8

B813—10: Specification for Liquid and Paste Fluxes for Soldering of Copper and Copper Alloy Tube
605.12.3, 605.13.6, 705.5.3, 705.6.1, 705.7.3

B828—02 (2010): Practice for Making Capillary Joints by Soldering of Copper and Copper Alloy Tube and Fittings
605.12.3, 605.13.6, 705.5.3, 705.6.1, 705.17.3

C4—04 (2014): Specification for Clay Drain Tile and Perforated Clay Drain Tile
Table 702.3, Table 1102.4, Table 1102.5

C14—15a: Specification for Nonreinforced Concrete Sewer, Storm Drain and Culvert Pipe
Table 702.3, Table 1102.4

C76—15a: Specification for Reinforced Concrete Culvert, Storm Drain and Sewer Pipe
Table 702.3, Table 1102.4

C425—04 (2013): Specification for Compression Joints for Vitrified Clay Pipe and Fittings
705.11, 705.16

C443—12: Specification for Joints for Concrete Pipe and Manholes, Using Rubber Gaskets
705.4, 705.16

C564—14: Specification for Rubber Gaskets for Cast-iron Soil Pipe and Fittings
705.3.2, 705.3.3, 705.16

C700—13: Specification for Vitrified Clay Pipe, Extra Strength, Standard Strength, and Perforated
Table 702.3, Table 702.4, Table 1102.4, Table 1102.5

C1053—00 (2010): Specification for Borosilicate Glass Pipe and Fittings for Drain, Waste, and Vent (DWV) Applications
Table 702.1, Table 702.4

C1173—10 (2014): Specification for Flexible Transition Couplings for Underground Piping System
705.2.1, 705.5, 705.10.1, 705.11, 705.12.2, 705.16

ASTM—continued

C1277—15: Specification for Shielded Coupling Joining Hubless Cast-iron Soil Pipe and Fittings
    705.3.3

C1440—08 (2013): Specification for Thermoplastic Elastomeric (TPE) Gasket Materials for Drain, Waste, and Vent (DWV), Sewer, Sanitary and Storm Plumbing Systems
    705.16

C1460—2012: Specification for Shielded Transition Couplings for Use with Dissimilar DWV Pipe and Fittings Above Ground
    705.16

C1461—08 (2013): Specification for Mechanical Couplings Using Thermoplastic Elastomeric (TPE) Gaskets for Joining Drain, Waste and Vent (DWV) Sewer, Sanitary and Storm Plumbing Systems for Above and Below Ground Use
    705.19

C1540—15: Specification for Heavy Duty Shielded Couplings Joining Hubless Cast-iron Soil Pipe and Fittings
    705.3.3

C1563—08 (2013): Standard Test Method for Gaskets for Use in Connection with Hub and Spigot Cast Iron Soil Pipe and Fittings for Sanitary Drain, Waste, Vent and Storm Piping Applications
    705.3.2

D1253—14: Standard Test Method For Residual Chlorine in Water
    1301.2.1, 1303.9

D1527—99 (2005): Specification for Acrylonitrile-Butadiene-Styrene (ABS) Plastic Pipe, Schedules 40 and 80
    Table 605.3

D1785—15: Specification for Poly (Vinyl Chloride) (PVC) Plastic Pipe, Schedules 40, 80 and 120
    Table 605.3

D2235—04 (2011): Specification for Solvent Cement for Acrylonitrile-Butadiene-Styrene (ABS) Plastic Pipe and Fittings
    605.10.2, 705.2.2, 705.7.2

D2239—12a: Specification for Polyethylene (PE) Plastic Pipe (SIDR-PR) Based on Controlled Inside Diameter
    Table 605.3

D2241—15: Specification for Poly (Vinyl Chloride) (PVC) Pressure-rated Pipe (SDR-Series)
    Table 605.3

D2282—99(2005): Specification for Acrylonitrile-Butadiene-Styrene (ABS) Plastic Pipe (SDR-PR)
    Table 605.3

D2464—15: Specification for Threaded Poly (Vinyl Chloride) (PVC) Plastic Pipe Fittings, Schedule 80
    Table 605.5, 605.21.3

D2466—15: Specification for Poly (Vinyl Chloride) (PVC) Plastic Pipe Fittings, Schedule 40
    Table 605.5

D2467—15: Specification for Poly (Vinyl Chloride) (PVC) Plastic Pipe Fittings, Schedule 80
    Table 605.5

D2468—96a: Specification for Acrylonitrile-Butadiene-Styrene (ABS) Plastic Pipe Fittings, Schedule 40
    Table 605.5

D2564—12: Specification for Solvent Cements for Poly (Vinyl Chloride) (PVC) Plastic Piping Systems
    605.21.3, 705.10.2, 705.14.2

D2609—15: Specification for Plastic Insert Fittings for Polyethylene (PE) Plastic Pipe
    Table 605.5

D2657—07: Practice for Heat Fusion-joining of Polyolefin Pipe and Fitting Waste, and Vent Pipe and Fittings
    605.18.2, 705.12.1

D2661—14: Specification for Acrylonitrile-Butadiene-Styrene (ABS) Schedule 40 Plastic Drain, Waste, and Vent Pipe and Fittings
    Table 702.1, Table 702.2, Table 702.3, Table 702.4, 705.2.2, 705.7.2, Table 1102.4, Table 1102.7

D2665—14: Specification for Poly (Vinyl Chloride) (PVC) Plastic Drain, Waste, and Vent Pipe and Fittings
    Table 702.1, Table 702.2, Table 702.3, Table 702.4, Table 1102.4, Table 1102.7

D2672—14: Specification for Joints for IPS PVC Pipe Using Solvent Cement
    Table 605.3

D2683—14: Standard Specification for Socket-type Polyethylene Fittings for Outside Diameter-controlled Polyethylene Pipe and Tubing
    Table 605.5, Table 702.4, 716.5

REFERENCED STANDARDS

ASTM—continued

D2729—11: Specification for Poly (Vinyl Chloride) (PVC) Sewer Pipe and Fittings
    Table 1102.5, Table 1403.2

D2737—2012a: Specification for Polyethylene (PE) Plastic Tubing
    Table 605.3

D2751—05: Specification for Acrylonitrile-Butadiene-Styrene (ABS) Sewer Pipe and Fittings
    Table 702.3, Table 702.4, Table 1102.7

D2846/D2846M—14: Specification for Chlorinated Poly (Vinyl Chloride) (CPVC) Plastic Hot- and Cold-Water Distribution Systems
    Table 605.3, Table 605.4, Table 605.5, 605.14.2, 605.15.2

D2855—96 (2010): Standard Practice for Making Solvent-cemented Joints with Poly (Vinyl Chloride) (PVC) Pipe and Fittings
    605.21.3, 705.10.2

D2949—10: Specification for 3.25-in. Outside Diameter Poly (Vinyl Chloride) (PVC) Plastic Drain, Waste, and Vent Pipe and Fittings
    Table 702.1, Table 702.2, Table 702.3, Table 702.4, Table 1102.7

D3034—14a: Specification for Type PSM Poly (Vinyl Chloride) (PVC) Sewer Pipe and Fittings
    Table 702.3, Table 702.4, Table 1102.4, Table 1102.5, Table 1102.7

D3035—15: Standard Specification for Polyethylene (PE) Plastic Pipe (DR-PR) Based on Controlled Outside Diameter
    Table 605.3

D3138—04 (2011): Standard Specification for Solvent Cements for Transition Joints Between Acrylonitrile-Butadiene-Styrene (ABS) and Poly (Vinyl Chloride) (PVC) Non-pressure Piping Components
    Table 705.16.4

D3139—98 (2011): Specification for Joints for Plastic Pressure Pipes Using Flexible Elastomeric Seals
    605.10.1, 605.22.1

D3212—07 (2013): Specification for Joints for Drain and Sewer Plastic Pipes Using Flexible Elastomeric Seals
    705.2.1, 705.10.1, 705.12.2, 705.15

D3261—12e1: Specification for Butt Heat Fusion Polyethylene (PE) Plastic Fittings for Polyethylene (PE) Plastic Pipe and Tubing
    Table 605.5

D3311—11: Specification for Drain, Waste and Vent (DWV) Plastic Fittings Patterns
    Table 1102.7

D4068—15: Specification for Chlorinated Polyethylene (CPE) Sheeting for Concealed Water-containment Membrane
    422.5.2.2

D4551—12: Specification for Poly (Vinyl Chloride) (PVC) Plastic Flexible Concealed Water-containment Membrane
    421.5.2.1

E2635—14: Standard Practice for Water Conservation Through In-Situ Water Reclamation
    1302.1

E2727—10e1: Standard Practice for the Assessment of Rainwater Quality
    1303.15.9

F405—05: Specification for Corrugated Polyethylene (PE) Pipe and Fittings
    Table 1102.5, Table 1403.2

F409—12: Specification for Thermoplastic Accessible and Replaceable Plastic Tube and Tubular Fittings
    412.1.2, Table 1102.7

F437—15: Specification for Threaded Chlorinated Poly (Vinyl Chloride) (CPVC) Plastic Pipe Fittings, Schedule 80
    Table 605.5

F438—15: Specification for Socket-type Chlorinated Poly (Vinyl Chloride) (CPVC) Plastic Pipe Fittings, Schedule 40
    Table 605.5

F439—13: Standard Specification for Chlorinated Poly (Vinyl Chloride) (CPVC) Plastic Pipe Fittings, Schedule 80
    Table 605.5

F441/F441M—15: Specification for Chlorinated Poly (Vinyl Chloride) (CPVC) Plastic Pipe, Schedules 40 and 80
    Table 605.3, Table 605.4

F442/F442M—13e1: Specification for Chlorinated Poly (Vinyl Chloride) (CPVC) Plastic Pipe (SDR-PR)
    Table 605.3, Table 605.4

F477—14: Specification for Elastomeric Seals (Gaskets) for Joining Plastic Pipe
    605.23, 705.16

REFERENCED STANDARDS

ASTM—continued

F493—14: Specification for Solvent Cements for Chlorinated Poly (Vinyl Chloride) (CPVC) Plastic Pipe and Fittings
    605.14.2, 605.15.2

F628—12e1: Specification for Acrylonitrile-Butadiene-Styrene (ABS) Schedule 40 Plastic Drain, Waste, and Vent Pipe with a Cellular Core
    Table 702.1, Table 702.2, Table 702.3, Table 702.4, 705.2.2, Table 1102.4, Table 1102.7

F656—15: Specification for Primers for Use in Solvent Cement Joints of Poly (Vinyl Chloride) (PVC) Plastic Pipe and Fittings
    605.21.3, 705.10.2, 705.14.2

F667—12: Standard Specification for 3 through 24 in. Corrugated Polyethylene Pipe and Fittings
    Table 1102.4, Table 1102.5

F714—2013: Specification for Polyethylene (PE) Plastic Pipe (SDR-PR) Based on Outside Diameter
    Table 702.3, 717.4

F876—15a: Specification for Cross-linked Polyethylene (PEX) Tubing
    Table 605.3, Table 605.4

F877—11a: Specification for Cross-linked Polyethylene (PEX) Hot- and Cold-water Distribution Systems
    Table 605.5

F891—10: Specification for Coextruded Poly (Vinyl Chloride) (PVC) Plastic Pipe with a Cellular Core
    Table 702.1 Table 702.2, Table 702.3, Table 1102.4, Table 1102.5, Table 1102.7

F1055—13: Standard Specification for Electrofusion Type Polyethylene Fittings for Outside Diameter Controlled Polyethylene and Cross-linked Polyethylene Pipe and Tubing
    Table 605.5

F1281—11: Specification for Cross-linked Polyethylene/Aluminum/ Cross-linked Polyethylene (PEX-AL-PEX) Pressure Pipe
    Table 605.3, Table 605.4, Table 605.5, 605.20.1

F1282—10: Specification for Polyethylene/Aluminum/Polyethylene (PE-AL-PE) Composite Pressure Pipe
    Table 605.3, Table 605.4, Table 605.5, 605.20.1

F1412—09: Specification for Polyolefin Pipe and Fittings for Corrosive Waste Drainage
    Table 702.1, Table 702.2, Table 702.4, 705.13.1, 901.3

F1476—07 (2013): Standard Specification for Performance of Gasketed Mechanical Couplings for Use in Piping Applications
    Table 605.5, 605.13.3, 605.17.3, 605.22.2, 605.22.3

F1488—14: Specification for Coextruded Composite Pipe
    Table 702.1, Table 702.2, Table 702.3, Table 1102.4, Table 1403.2

F1548—01 (2012): Standard Specification for the Performance of Fittings for Use with Gasketed Mechanical Couplings Used in Piping Applications
    Table 605.5

F1673—10: Standard Specification for Polyvinylidene Fluoride (PVDF) Corrosive Waste Drainage Systems
    Table 702.1, Table 702.2, Table 702.3, Table 702.4, 705.14.1

F1807—15: Specification for Metal Insert Fittings Utilizing a Copper Crimp Ring for SDR9 Cross-linked Polyethylene (PEX) Tubing and SDR9 Polyethylene of Raised Temperature (PE-RT) Tubing
    Table 605.5

F1866—13: Specification for Poly (Vinyl Chloride) (PVC) Plastic Schedule 40 Drainage and DWV Fabricated Fittings
    Table 702.4, Table 1102.7

F1960—15: Specification for Cold Expansion Fittings with PEX Reinforcing Rings for Use with Cross-linked Polyethylene (PEX) Tubing
    Table 605.5

F1970—12e1: Special Engineered Fittings, Appurtenances or Valves for Use in Poly (Vinyl Chloride) (PVC) OR Chlorinated Poly (Vinyl Chloride) (CPVC) Systems
    Table 605.7

F1974—09 (2015): Specification for Metal Insert Fittings for Polyethylene/Aluminum/Polyethylene and Cross-linked Polyethylene/ Aluminum/Cross-linked Polyethylene Composite Pressure Pipe
    Table 605.5, 605.20.1

F1986—01 (2011): Specification for Multilayer Pipe, Type 2, Compression Fittings and Compression Joints for Hot and Cold Drinking Water Systems
    Table 605.3, Table 605.4, Table 605.5

REFERENCED STANDARDS

ASTM—continued

F2080—15: Specifications for Cold-expansion Fittings with Metal Compression-sleeves for Cross-linked Polyethylene (PEX) Pipe
    Table 605.5

F2098—08: Standard Specification for Stainless Steel Clamps for Securing SDR9 Cross-linked Polyethylene (PEX) Tubing to Metal Insert and Plastic Fittings
    Table 605.5

F2159—14: Specification for Plastic Insert Fittings Utilizing a Copper Crimp Ring for SDR9 Cross-linked Polyethylene (PEX) Tubing and SDR9 Polyethylene of Raised Temperature (PE-RT) Tubing
    Table 605.5

F2262—09: Specification for Cross-linked Polyethylene/Aluminum/Cross-linked Polyethylene Tubing OD Controlled SDR9
    Table 605.3, Table 605.4

F2306/F2306M—14e1: 12" to 60" Annular Corrugated Profile-wall Polyethylene (PE) Pipe and Fittings for Gravity Flow Storm Sewer and Subsurface Drainage Applications
    Table 1102.4, Table 1102.7

F2389—15: Specification for Pressure-rated Polypropylene (PP) Piping Systems
    Table 605.3, Table 605.4, Table 605.5, Table 605.7, 605.19.1

F2434—14: Standard Specification for Plastic Insert Fittings Utilizing a Copper Crimp Ring for SDR9 Cross-linked Polyethylene (PEX) Tubing and SDR9 Cross-linked Polyethylene/Aluminum/ Cross-linked Polyethylene (PEX AL-PEX) Tubing
    Table 605.5

F2648/F2648M—13: Standard Specification for 2 to 60 inch [50 to 1500 mm] Annular Corrugated Profile Wall Polyethylene (PE) Pipe and Fittings for Land Drainage Applications
    Table 1102.4

F2735—09: Standard Specification for Plastic Insert Fittings for SDR9 Cross-linked Polyethylene (PEX) and Polyethylene of Raised Temperature (PE-RT) Tubing
    Table 605.5

F2736—13e1: Standard Specification for 6 to 30 in. [152 to 762 mm] Polypropylene (PP) Corrugated Single Wall Pipe and Double Wall Pipe
    Table 702.3

F2764/F2764M—11ae2: Standard Specification for 30 to 60 in. [750 to 1500 mm] Polypropylene (PP) Triple Wall Pipe and Fittings for Non-pressure Sanitary Sewer Applications
    Table 702.3

F2769—14: Polyethylene or Raised Temperature (PE-RT) Plastic Hot- and Cold-water Tubing and Distribution Systems
    Table 605.3, Table 605.4, Table 605.5

F2831—12: Standard Practice for Internal Non Structural Epoxy Barrier Coating Material Used in Rehabilitation of Metallic Pressurized Piping Systems
    601.5

F2855—12: Standard Specification for Chlorinated Poly (Vinyl Chloride)/Aluminum/Chlorinated Poly (Vinyl Chloride) (CPVC/AL/CPVC) Composite Pressure Tubing
    Table 605.3, Table 605.4

F2881—11: Standard Specification for 12 to 60 in. [300 to 1500 mm] Polypropylene (PP) Dual Wall Pipe and Fittings for Non-pressure Storm Sewer Applications
    Table 1102.4

AWS

American Welding Society
8669 NW 36 Street, #130
Miami, FL 33166

A5.8M/A5.8—2011: Specifications for Filler Metals for Brazing and Braze Welding
    605.12.1, 605.13.1, 705.5.1, 705.6.1

AWWA

American Water Works Association
6666 West Quincy Avenue
Denver, CO 80235

C104/A21.4—13: Cement-mortar Lining for Ductile-iron Pipe and Fittings
    605.3, 605.5

REFERENCED STANDARDS

AWWA—continued

C110/A21.10—12: Ductile-iron and Gray-iron Fittings
Table 605.5, Table 702.4, Table 1102.7

C111/A21.11—12: Rubber-gasket Joints for Ductile-iron Pressure Pipe and Fittings
605.11

C115/A21.15—11: Flanged Ductile-iron Pipe with Ductile-iron or Gray-iron Threaded Flanges
Table 605.3, Table 605.4

C151/A21.51—09: Ductile-iron Pipe, Centrifugally Cast for Water
Table 605.3, Table 605.4

C153—00/A21.53—11: Ductile-iron Compact Fittings for Water Service
Table 605.5

C500—09: Standard for Metal-seated Gate Valves for Water Supply Service
Table 605.7

C504—10: Standard for Rubber-Seated Butterfly Valves
Table 605.7

C507—15: Standard for Ball Valves, 6 In. Through 60 in. (150 mm through 1,500 mm).
Table 605.7

C510—07: Double Check Valve Backflow Prevention Assembly
Table 608.1, 608.14.7

C511—07: Reduced-pressure Principle Backflow Prevention Assembly
Table 608.1, 608.14.2, 608.17.2

C651—14: Disinfecting Water Mains
610.1

C652—11: Disinfection of Water-storage Facilities
610.1

C901—16: Polyethylene (PE) Pressure Pipe and Tubing $3/4$ inch (19 mm) Through 3 inch (76 mm) for Water Service
Table 605.3

C904—16: Cross-linked Polyethylene (PEX) Pressure Tubing $1/2$ inch (13 mm) Through 3 inch (76 mm) for Water Service
Table 605.3

CISPI

Cast Iron Soil Pipe Institute
2401 Fieldcrest Dr.
Mundelein, IL 60060

301—12: Specification for Hubless Cast-iron Soil Pipe and Fittings for Sanitary and Storm Drain, Waste and Vent Piping Applications
Table 702.1, Table 702.2, Table 702.3, Table 702.4, 708.7, Table 1102.4, Table 1102.5, Table 1102.7

310—12: Specification for Coupling for Use in Connection with Hubless Cast-iron Soil Pipe and Fittings for Sanitary and Storm Drain, Waste and Vent Piping Applications
705.3.3

CSA

CSA Group
8501 East Pleasant Valley Road
Cleveland, OH 44131-5516

A257.1M—14: Non-reinforced Circular Concrete Culvert, Storm Drain, Sewer Pipe and Fittings
Table 702.3, Table 1102.4

A257.2M—14: Reinforced Circular Concrete Culvert, Storm Drain, Sewer Pipe and Fittings
Table 702.3, Table 1102.4

A257.3M—14: Joints for Circular Concrete Sewer and Culvert Pipe, Manhole Sections and Fittings Using Rubber Gaskets
705.5, 705.16

ASME A112.18.1—2017/CSA B125.1—17: Plumbing Supply Fittings
412.1, 412.2, 412.3, 412.4, 412.6, 412.8, Table 605.7, 607.4, 608.2

ASME A112.18.2—2015/CSA B125.2—2015: Plumbing Waste Fittings
412.1.2

REFERENCED STANDARDS

CSA—continued

ASME A112.19.1—2013/CSA B45.2—2013: Enameled Cast-iron and Enameled Steel Plumbing Fixtures
    407.1, 410.1, 418.1, 419.1, 422.1

ASME A112.19.2—2013/B45.1—2013: Ceramic Plumbing Fixtures
    401.2, 405.10, 407.1, 408.1, 410.1, 418.1, 419.1, 421.1, 422.1, 424.1, 425.1

ASME A112.19.3—2008/CSA B45.4—08 (R2013): Stainless-steel Plumbing Fixtures
    405.9, 407.1, 418.1, 419.1, 425.1

ASME A112.19.5—2017/CSA B45.15—17: Flush Valves and Spuds for Water Closets, Urinals and Tanks
    415.4

ASME A112.19.7—2017/CSA B45.10—17: Hydromassage Bathtub Systems
    426.1, 426.4

CSA B45.5—17/IAPMO Z124—2017: Plastic Plumbing Fixtures
    407.1, 418.1, 419.1, 419.2, 421.1, 424.1, 425.1

ASME A112.3.4—2013/CSA B45.9—13: Macerating Systems and Related Components
    405.5, 712.4.1

ASSE 1002—2015/ASME A112.1002—2015/CSA B125.12—2015: Anti-Siphon Fill Valves
    415.3.1, Table 608.1

ASSE 1016/ASME A112.1016/CSA B125.16—2017: Performance Requirements for Automatic Compensating Valves for Individual Showers and Tub/Shower Combinations
    412.3, 412.4, 607.4

ASSE 1037—2015/ASME A112.1037—2015/CSA B125.37—15: Pressurized Flushing Devices for Plumbing Fixtures
    415.2

ASSE 1070—2015/ASME A112.1070—2015/CSA B125.1070—2015: Water Temperature Limiting Devices
    408.3, 412.5, 412.10, 419.5, 423.3, 607.1.2

B64.1.1—16: Vacuum Breakers, Atmospheric Type (AVB)
    415.2, Table 608.1, 608.14.6, 608.17.4.1

B64.1.2—16: Pressure Vacuum Breakers, (PVB)
    Table 608.1, 608.14.5

B64.1.3—16: Spill Resistant Pressure Vacuum Breakers (SRPVB)
    Table 608.1, 608.14.5

B64.2—16: Vacuum Breakers, Hose Connection Type (HCVB)
    Table 608.1, 608.14.6

B64.2.1—16: Vacuum Breakers, Hose Connection (HCVB) with Manual Draining Feature
    Table 608.1, 608.14.6

B64.2.1.1—16: Hose Connection Dual Check Vacuum Breakers (HCDVB)
    Table 608.1, 608.14.6

B64.2.2—16: Vacuum Breakers, Hose Connection Type (HCVB) with Automatic Draining Feature
    Table 608.1, 608.14.6

B64.3—16: Backflow Preventers, Dual Check Valve Type with Atmospheric Port (DCAP)
    Table 608.1, 608.14.3, 608.17.2

B64.4—16: Backflow Preventers, Reduced Pressure Principle Type (RP)
    Table 608.1, 608.14.2, 608.17.2

B64.4.1—16: Reduced Pressure Principle for Fire Sprinklers (RPF)
    Table 608.1, 608.14.2

B64.5—16: Double Check Backflow Preventers (DCVA)
    Table 608.1, 608.14.7

B64.5.1—16: Double Check Valve Backflow Preventer for Fire Systems (DCVAF)
    Table 608.1, 608.14.7

B64.6—16: Dual Check Valve (DuC) Backflow Preventers
    605.3.1, Table 608.1, 608.14.9

B64.7—16: Laboratory Faucet Vacuum Breakers (LFVB)
    Table 608.1, 608.14.6

CSA—continued

B64.10—16: Manual for the Selection and Installation of Backflow Prevention Devices
    312.10.2

B64.10.1—11: Maintenance and Field Testing of Backflow Preventers
    312.10.2

B79—08(R2013): Commercial and Residential Drains and Cleanouts
    413.1

B125.3—2012: Plumbing Fittings
    408.3, 412.4, 412.5, 415.2, 415.3.1, 419.5, 423.3, Table 605.7, Table 608.1

B137.1—16: Polyethylene (PE) Pipe, Tubing and Fittings for Cold-water Pressure Services
    Table 605.3, Table 605.5

B137.2—16: Polyvinylchloride, PVC, Injection-moulded Gasketed Fittings for Pressure Applications
    Table 605.5

B137.3—16: Rigid Poly (Vinyl Chloride) (PVC) Pipe for Pressure Applications
    Table 605.3, Table 605.5, 605.21.3, 705.10.2, 705.14.2

B137.5—16: Cross-linked Polyethylene (PEX) Tubing Systems for Pressure Applications
    Table 605.3, Table 605.4, Table 605.5

B137.6—16: CPVC Pipe, Tubing and Fittings for Hot- and Cold-water Distribution Systems
    Table 605.3, Table 605.4, Table 605.5

B137.9—16: Polyethylene Aluminum/Polyethylene (PE-AL-PE) Composite Pressure-pipe Systems
    Table 605.3, Table 605.5, 605.20.1

B137.10—16: Cross-linked Polyethylene/Aluminum/Cross-linked Polyethylene (PEX-AL-PEX) Composite Pressure-pipe Systems
    Table 605.3, Table 605.4, Table 605.5, 605.20.1

B137.11—16: Polypropylene (PP-R) Pipe and Fittings for Pressure Applications
    Table 605.3, Table 605.4, Table 605.5

B137.18—13: Polyethylene of Raised Temperature Resistance (PE-RT) Tubing Systems for Pressure Applications
    Table 605.3, Table 605.4, Table 605.5

B181.1—15: Acrylonitrile-Butadiene-Styrene ABS Drain, Waste and Vent Pipe and Pipe Fittings
    Table 702.1, Table 702.2, Table 702.3, Table 702.4, 705.2.2, 714.2, Table 1102.4, Table 1102.7

B181.2—15: Polyvinylchloride PVC and Chlorinated Polyvinylchloride (CPVC) Drain, Waste, and Vent Pipe and Pipe Fittings
    Table 702.1 Table 702.2, 705.10.2, 705.14.2, 714.2

B181.3—15: Polyolefin and Polyvinylidene Fluoride (PVDF) Laboratory Drainage Systems
    Table 702.1, Table 702.2, Table 702.3, Table 702.4, 705.13.1

B182.1—11: Plastic Drain and Sewer Pipe and Pipe Fittings
    705.10.2, 705.14.2, Table 1102.4, Table 1102.5

B182.2—11: PSM Type Polyvinylchloride PVC Sewer Pipe and Fittings
    Table 702.3, Table 1102.4, Table 1102.5

B182.4—15: Profile Polyvinylchloride PVC Sewer Pipe and Fittings
    Table 702.3, Table 1102.4, Table 1102.5

B182.6—15: Profile Polyethylene (PE) Sewer Pipe and Fittings for Leak-proof Sewer Applications
    Table 1102.5

B182.8—15: Profile Polyethylene (PE) Storm Sewer and Drainage Pipe and Fittings
    Table 1102.5

B182.13—11: Profile Polypropylene (PP) Sewer Pipe and Fittings for Leak-proof Sewer Applications
    Table 702.3, Table 1102.4

B356—10: Water Pressure Reducing Valves for Domestic Water Systems
    604.8

B481.1—12: Testing and Rating of Grease Interceptors Using Lard
    1003.3.4

B481.3—12: Sizing, Selection, Location and Installation of Grease Interceptors
    1003.3.5

B483.1—07(R2012): Drinking Water Treatment Units
    611.1, 611.2

REFERENCED STANDARDS

CSA—continued
B602—15: Mechanical Couplings for Drain, Waste and Vent Pipe and Sewer Pipe
    705.2.1, 705.3.3, 705.5, 705.10.1, 705.11, 705.12.2, 705.16

IAPMO

IAPMO Group
4755 E. Philadelphia Street
Ontario, CA 91761 USA

Z1001—2014: Prefabricated Gravity Grease Interceptors
    1003.3.7

CSA B45.5—17/IAPMO Z124—2017: Plastic Plumbing Fixtures
    407.1, 418.1, 419.1, 419.2, 421.1, 424.1, 425.1

IAPMO/ANSI Z1157—2014: Ball Valves
    Table 605.7

ICC

International Code Council, Inc.
500 New Jersey Ave, NW
6th Floor
Washington, DC 20001

A117.1—2009: Accessible and Usable Buildings and Facilities
    404.2, 410.3

IBC—18: International Building Code®
    201.3, 202, 307.1, 307.2, 307.3, 308.2, 309.1, 309.2, 310.1, 310.3, 315.1, 403.1, Table 403.1, 403.1.2, 403.3.1, 403.4, 404.1, 407.3, 421.6, 502.4, 606.5.2, 1106.5, 1301.9.3, 1303.6, 1402.1

IEBC—18: International Existing Building Code
    102.2.1

IECC—18: International Energy Conservation Code®
    313.1, 607.2.1, 607.5

IFC—18: International Fire Code®
    201.3, 1201.1, 1301.1.1

IFGC—18: International Fuel Gas Code®
    101.2, 201.3, 502.1, 502.1.1

IMC—18: International Mechanical Code®
    201.3, 307.6, 310.1, 502.1, 502.1.1, 612.1, 1202.1

IPSDC—18: International Private Sewage Disposal Code®
    701.2

IRC—18: International Residential Code®
    101.2

ICC 900/SRCC 300—2015: Solar Thermal System Standard
    502.1

ISEA

International Safety Equipment Association
1901 N. Moore Street, Suite 808
Arlington, VA 22209

ANSI/ISEA Z358.1—2014: Emergency Eyewash and Shower Equipment
    411.1

MSS

Manufacturers Standardization Society of the Valve and Fittings Industry, Inc.
127 Park St. NE
Vienna, VA 22180-4602

SP-67—2011: Butterfly Valves
    Table 605.7

MSS—continued

SP-70—2013: Gray Iron Gate Valves, Flanged and Threaded Ends
Table 605.7

SP-71—2013: Gray Iron Swing Check Valves, Flanged and Threaded Ends
Table 605.7

SP-72—2010a: Ball Valves with Flanged or Butt-welding Ends for General Service
Table 605.7

SP-78—2013: Cast Iron Plug Valves, Flanged and Threaded Ends
Table 605.7

SP-80—2013: Bronze Gate, Globe, Angle and Check Valves
Table 605.7

SP-110—2010a: Ball Valves, Threaded, Socket Welding, Solder Joint, Grooved and Flared Ends
Table 605.7

SP-122—2012: Plastic Industrial Ball Valves
Table 605.7

SP-139—2014: Copper Alloy Gate, Globe, Angle and Check Valves for Low Pressure/Low Temperature Plumbing Applications
Table 605.7

NFPA

National Fire Protection Association
1 Batterymarch Park
Quincy, MA 02169-7471

51—18: Design and Installation of Oxygen-fuel Gas Systems for Welding, Cutting and Allied Processes
1203.1

55—16: Compressed Gases and Cryogenic Fluids Code
1203.1

70—17: National Electric Code
502.1, 504.3, 1113.1.3

99—18: Health Care Facilities Code
1202.1

NGWA

National Ground Water Association
601 Dempsey Road
Westerville, OH 43081

ANSI/NGWA 01—14: Water Well Construction Standard
602.3.1

NSF

NSF International
789 N. Dixboro Road
P.O. Box 130140
Ann Arbor, MI 48105

3—2012: Commercial Warewashing Equipment
409.1

14—2015: Plastic Piping System Components and Related Materials
303.3, 611.3

18—2012: Manual Food and Beverage Dispensing Equipment
426.1

42—2015: Drinking Water Treatment Units-Aesthetic Effects
611.1, 611.3

44—2015: Residential Cation Exchange Water Softeners
611.1, 611.3

50—2015: Equipment for Swimming Pools, Spas, Hot Tubs and Other Recreational Facilities
1302.8.1

REFERENCED STANDARDS

NSF—continued

53—2015: Drinking Water Treatment Units—Health Effects
611.1, 611.3

58—2015: Reverse Osmosis Drinking Water Treatment Systems
611.1, 611.2, 611.3

61—2015: Drinking Water System Components—Health Effects
410.1, 412.1, 605.3, 605.4, 605.5, 605.7, 608.12, 611.3

62—2015: Drinking Water Distillation Systems
611.1

184—2014: Residential Dishwashers
409.1

350—2014: Onsite Residential and Commercial Water Reuse Treatment Systems
1302.6.1

359—2011: Valves for Cross-linked Polyethylene (PEX) Water Distribution Tubing Systems
Table 605.7

372—2011: Drinking Water Systems Components—Lead Content
605.2.1

PDI

Plumbing and Drainage Institute
800 Turnpike Street, Suite 300
North Andover, MA 01845

PDI G101 (2012): Testing and Rating Procedure for Grease Interceptors with Appendix of Sizing and Installation Data
1003.3.5

PDI G102 (2009): Testing and Certification for Grease Interceptors with Fog Sensing and Alarm Devices
1003.3.5

PSAI

Portable Sanitation Association International
2626 E. 82nd Street, Suite 175
Bloomington, MN 55425

PSAI/ANSI Z4.3—16: Minimum Requirements for Nonsewered Waste-disposal Systems
311.1

UL

UL LLC
333 Pfingsten Road
Northbrook, IL 60062-2096

399—2008: Drinking-Water Coolers—with revisions through October 2013
410.1

430—2009: Waste Disposers—with revisions through September 2015
416.1

508—99: Industrial Control Equipment—with revisions through October 2013
314.2.3

1795—2009: Hydromassage Bathtubs—with revisions through January 2015
426.1

APPENDIX A

PLUMBING PERMIT FEE SCHEDULE

The provisions contained in this appendix are not mandatory unless specifically referenced in the adopting ordinance.

User note:

About this appendix: *Appendix A provides an example of a permit fee schedule that can be used by a jurisdiction.*

Permit Issuance

1. For issuing each permit. . . . . $ _____

2. For issuing each supplemental permit . . . . _____

Unit Fee Schedule

1. For each plumbing fixture or trap or set of fixtures on one trap (including water, drainage piping and backflow protection thereof). . . . . _____

2. For each building sewer and each trailer park sewer . . . . _____

3. Rainwater systems—per drain (inside building) . . . . _____

4. For each cesspool (where permitted) . . . . _____

5. For each private sewage disposal system . . . . _____

6. For each water heater and/or vent. . . . . _____

7. For each industrial waste pretreatment interceptor including its trap and vent, excepting kitchen-type grease interceptors functioning as fixture traps . . . . _____

8. For installation, alteration or repair of water-piping and/or water-treating equipment, each . . . . _____

9. For repair or alteration of drainage or vent piping, each fixture . . . . _____

10. For each lawn sprinkler system on any one meter including backflow protection devices therefor . . . . _____

11. For atmospheric-type vacuum breakers not included in Item 2:

    1 to 5 . . . . _____

    over 5, each . . . . _____

12. For each backflow protective device other than atmospheric-type vacuum breakers:

    2 inches (51 mm) and smaller . . . . _____

    Over 2 inches (51 mm) . . . . _____

Other Inspections and Fees

1. Inspections outside of normal business hours (minimum charge 2 hours). . . . . _____ per hour

2. Reinspection fee assessed under provisions of Section 107.4.3. . . . . _____ each

3. Inspections for which no fee is specifically indicated (minimum charge one-half hour) . . . . _____ per hour

4. Additional plan review required by changes, additions or revisions to approved plans (minimum charge one-half hour). . . . . _____ per hour

APPENDIX B

RATES OF RAINFALL FOR VARIOUS CITIES

This appendix is informative and is not part of the code.

User note:

About this appendix: The design of storm water drainage systems (Chapter 11) requires determination of the design rainfall event for the geographic location of the building. Appendix B provides a list of major cities in the U.S. along with the 1-hour duration, 100-year occurrence rainfall rate.

Rainfall rates, in inches per hour, are based on a storm of 1-hour duration and a 100-year return period. The rainfall rates shown in the appendix are derived from Figure 1106.1.

Alabama:
- Birmingham 3.8
- Huntsville 3.6
- Mobile 4.6
- Montgomery 4.2

Alaska:
- Fairbanks 1.0
- Juneau 0.6

Arizona:
- Flagstaff 2.4
- Nogales 3.1
- Phoenix 2.5
- Yuma 1.6

Arkansas:
- Fort Smith 3.6
- Little Rock 3.7
- Texarkana 3.8

California:
- Barstow 1.4
- Crescent City 1.5
- Fresno 1.1
- Los Angeles 2.1
- Needles 1.6
- Placerville 1.5
- San Fernando 2.3
- San Francisco 1.5
- Yreka 1.4

Colorado:
- Craig 1.5
- Denver 2.4
- Durango 1.8
- Grand Junction 1.7
- Lamar 3.0
- Pueblo 2.5

Connecticut:
- Hartford 2.7
- New Haven 2.8
- Putnam 2.6

Delaware:
- Georgetown 3.0
- Wilmington 3.1

District of Columbia:
- Washington 3.2

Florida:
- Jacksonville 4.3
- Key West 4.3
- Miami 4.7
- Pensacola 4.6
- Tampa 4.5

Georgia:
- Atlanta 3.7
- Dalton 3.4
- Macon 3.9
- Savannah 4.3
- Thomasville 4.3

Hawaii:
- Hilo 6.2
- Honolulu 3.0
- Wailuku 3.0

Idaho:
- Boise 0.9
- Lewiston 1.1
- Pocatello 1.2

Illinois:
- Cairo 3.3
- Chicago 3.0
- Peoria 3.3
- Rockford 3.2
- Springfield 3.3

Indiana:
- Evansville 3.2
- Fort Wayne 2.9
- Indianapolis 3.1

Iowa:
- Davenport 3.3
- Des Moines 3.4
- Dubuque 3.3
- Sioux City 3.6

Kansas:
- Atwood 3.3
- Dodge City 3.3
- Topeka 3.7
- Wichita 3.7

Kentucky:
- Ashland 3.0
- Lexington 3.1
- Louisville 3.2
- Middlesboro 3.2
- Paducah 3.3

Louisiana:
- Alexandria 4.2
- Lake Providence 4.0
- New Orleans 4.8
- Shreveport 3.9

Maine:
- Bangor 2.2
- Houlton 2.1
- Portland 2.4

Maryland:
- Baltimore 3.2
- Hagerstown 2.8
- Oakland 2.7
- Salisbury 3.1

Massachusetts:
- Boston 2.5
- Pittsfield 2.8
- Worcester 2.7

Michigan:
- Alpena 2.5
- Detroit 2.7
- Grand Rapids 2.6
- Lansing 2.8
- Marquette 2.4
- Sault Ste. Marie 2.2

Minnesota:
- Duluth 2.8
- Grand Marais 2.3
- Minneapolis 3.1
- Moorhead 3.2
- Worthington 3.5

Mississippi:
- Biloxi 4.7
- Columbus 3.9
- Corinth 3.6
- Natchez 4.4
- Vicksburg 4.1

Missouri:
- Columbia 3.2
- Kansas City 3.6
- Springfield 3.4
- St. Louis 3.2

Montana:
- Ekalaka 2.5
- Havre 1.6
- Helena 1.5
- Kalispell 1.2
- Missoula 1.3

Nebraska:
- North Platte 3.3
- Omaha 3.8
- Scottsbluff 3.1
- Valentine 3.2

Nevada:
- Elko 1.0
- Ely 1.1
- Las Vegas 1.4
- Reno 1.1

New Hampshire:
- Berlin 2.5
- Concord 2.5
- Keene 2.4

New Jersey:
- Atlantic City 2.9
- Newark 3.1
- Trenton 3.1

New Mexico:
- Albuquerque 2.0
- Hobbs 3.0
- Raton 2.5
- Roswell 2.6
- Silver City 1.9

New York:
- Albany 2.5
- Binghamton 2.3
- Buffalo 2.3
- Kingston 2.7
- New York 3.0
- Rochester 2.2

North Carolina:
- Asheville 4.1
- Charlotte 3.7
- Greensboro 3.4
- Wilmington 4.2

APPENDIX B

North Dakota:
  Bismarck 2.8
  Devils Lake 2.9
  Fargo 3.1
  Williston. 2.6

Ohio:
  Cincinnati. 2.9
  Cleveland 2.6
  Columbus. 2.8
  Toledo 2.8

Oklahoma:
  Altus. 3.7
  Boise City 3.3
  Durant 3.8
  Oklahoma City. 3.8

Oregon:
  Baker 0.9
  Coos Bay 1.5
  Eugene 1.3
  Portland 1.2

Pennsylvania:
  Erie. 2.6
  Harrisburg 2.8
  Philadelphia 3.1
  Pittsburgh. 2.6
  Scranton 2.7

Rhode Island:
  Block Island 2.75
  Providence 2.6

South Carolina:
  Charleston 4.3
  Columbia 4.0
  Greenville. 4.1

South Dakota:
  Buffalo. 2.8
  Huron 3.3
  Pierre 3.1
  Rapid City 2.9
  Yankton 3.6

Tennessee:
  Chattanooga. 3.5
  Knoxville 3.2
  Memphis 3.7
  Nashville 3.3

Texas:
  Abilene. 3.6
  Amarillo. 3.5
  Brownsville 4.5
  Dallas 4.0
  Del Rio 4.0
  El Paso. 2.3
  Houston. 4.6
  Lubbock 3.3
  Odessa. 3.2
  Pecos 3.0
  San Antonio. 4.2

Utah:
  Brigham City. 1.2
  Roosevelt. 1.3
  Salt Lake City 1.3
  St. George 1.7

Vermont:
  Barre 2.3
  Bratteboro 2.7
  Burlington 2.1
  Rutland 2.5

Virginia:
  Bristol 2.7
  Charlottesville 2.8
  Lynchburg. 3.2
  Norfolk 3.4
  Richmond 3.3

Washington:
  Omak. 1.1
  Port Angeles 1.1
  Seattle 1.4
  Spokane. 1.0
  Yakima 1.1

West Virginia:
  Charleston. 2.8
  Morgantown 2.7

Wisconsin:
  Ashland. 2.5
  Eau Claire 2.9
  Green Bay. 2.6
  La Crosse. 3.1
  Madison. 3.0
  Milwaukee. 3.0

Wyoming:
  Cheyenne. 2.2
  Fort Bridger. 1.3
  Lander. 1.5
  New Castle 2.5
  Sheridan 1.7
  Yellowstone Park 1.4

APPENDIX C

STRUCTURAL SAFETY

The provisions contained in this appendix are not mandatory unless specifically referenced in the adopting ordinance.

User note:

About this appendix: *The installation of plumbing systems frequently requires piping to pass through building framing members. Appendix C provides the regulations for limits on the sizes and locations of holes that can be drilled or punched in various types of framing members.*

SECTION C101
CUTTING, NOTCHING AND BORING IN WOOD MEMBERS

[BS] C101.1 Joist notching. Notches on the ends of joists shall not exceed one-fourth the joist depth. Holes bored in joists shall not be within 2 inches (51 mm) of the top or bottom of the joist, and the diameter of any such hole shall not exceed one-third the depth of the joist. Notches in the top or bottom of joists shall not exceed one-sixth the depth and shall not be located in the middle third of the span.

[BS] C101.2 Stud cutting and notching. In exterior walls and bearing partitions, any wood stud is permitted to be cut or notched to a depth not exceeding 25 percent of its width. Cutting or notching of studs to a depth not greater than 40 percent of the width of the stud is permitted in nonbearing partitions supporting no loads other than the weight of the partition.

[BS] C101.3 Bored holes. The diameter of bored holes in wood studs shall not exceed 40 percent of the stud depth. The diameter of bored holes in wood studs shall not exceed 60 percent of the stud depth in nonbearing partitions. The diameter of bored holes in wood studs shall not exceed 60 percent of the stud depth in any wall where each stud is doubled, provided that not more than two such successive doubled studs are so bored. The edge of the bored hole shall not be closer than $^5/_8$ inch (15.9 mm) to the edge of the stud. Bored holes shall be not located at the same section of stud as a cut or notch.

[BS] C101.4 Cutting, notching and boring holes in structural steel framing. The cutting, notching and boring of holes in structural steel framing members shall be as prescribed by the registered design professional.

[BS] C101.5 Cutting, notching and boring holes in cold-formed steel framing. Flanges and lips of load-bearing cold-formed steel framing members shall not be cut or notched. Holes in webs of load-bearing cold-formed steel framing members shall be permitted along the centerline of the web of the framing member and shall not exceed the dimensional limitations, penetration spacing or minimum hole edge distance as prescribed by the registered design professional. Cutting, notching and boring holes of steel floor/roof decking shall be as prescribed by the registered design professional.

[BS] C101.6 Cutting, notching and boring holes in nonstructural cold-formed steel wall framing. Flanges and lips of nonstructural cold-formed steel wall studs shall not be cut or notched. Holes in webs of nonstructural cold-formed steel wall studs shall be permitted along the centerline of the web of the framing member, shall not exceed $1^1/_2$ inches (38 mm) in width or 4 inches (102 mm) in length, and the holes shall not be spaced less than 24 inches (610 mm) center to center from another hole or less than 10 inches (254 mm) from the bearing end.

APPENDIX D

DEGREE DAY AND DESIGN TEMPERATURES

This appendix is informative and is not part of the code.

User note:

About this appendix: This code refers to the 97.5 percent winter design temperature for directing the code user to requirements for design of the plumbing system. Appendix D provides such temperatures for many major cities in the United States.

**TABLE D101
DEGREE DAY AND DESIGN TEMPERATURES[a] FOR CITIES IN THE UNITED STATES**

STATE	STATION[b]	HEATING DEGREE DAYS (yearly total)	DESIGN TEMPERATURES			DEGREES NORTH LATITUDE[c]
			Winter	Summer		
			97$\frac{1}{2}$%	Dry bulb 2$\frac{1}{2}$%	Wet bulb 2$\frac{1}{2}$%	
AL	Birmingham	2,551	21	94	77	33°30'
	Huntsville	3,070	16	96	77	34°40'
	Mobile	1,560	29	93	79	30°40'
	Montgomery	2,291	25	95	79	32°20'
AK	Anchorage	10,864	-18	68	59	61°10'
	Fairbanks	14,279	-47	78	62	64°50'
	Juneau	9,075	1	70	59	58°20'
	Nome	14,171	-27	62	56	64°30'
AZ	Flagstaff	7,152	4	82	60	35°10'
	Phoenix	1,765	34	107	75	33°30'
	Tuscon	1,800	32	102	71	33°10'
	Yuma	974	39	109	78	32°40'
AR	Fort Smith	3,292	17	98	79	35°20'
	Little Rock	3,219	20	96	79	34°40'
	Texarkana	2,533	23	96	79	33°30'
CA	Fresno	2,611	30	100	71	36°50'
	Long Beach	1,803	43	80	69	33°50'
	Los Angeles	2,061	43	80	69	34°00'
	Los Angeles[d]	1,349	40	89	71	34°00'
	Oakland	2,870	36	80	64	37°40'
	Sacramento	2,502	32	98	71	38°30'
	San Diego	1,458	44	80	70	32°40'
	San Francisco	3,015	38	77	64	37°40'
	San Francisco[d]	3,001	40	71	62	37°50'
CO	Alamosa	8,529	-16	82	61	37°30'
	Colorado Springs	6,423	2	88	62	38°50'
	Denver	6,283	1	91	63	39°50'
	Grand Junction	5,641	7	94	63	39°10'
	Pueblo	5,462	0	95	66	38°20'
CT	Bridgeport	5,617	9	84	74	41°10'
	Hartford	6,235	7	88	75	41°50'
	New Haven	5,897	7	84	75	41°20'
DE	Wilmington	4,930	14	89	76	39°40'
DC	Washington	4,224	17	91	77	38°50'

(continued)

APPENDIX D

TABLE D101—continued
DEGREE DAY AND DESIGN TEMPERATURES[a] FOR CITIES IN THE UNITED STATES

STATE	STATION[b]	HEATING DEGREE DAYS (yearly total)	DESIGN TEMPERATURES			DEGREES NORTH LATITUDE[c]
			Winter	Summer		
			97½%	Dry bulb 2½%	Wet bulb 2½%	
FL	Daytona	879	35	90	79	29°10′
	Fort Myers	442	44	92	79	26°40′
	Jacksonville	1,239	32	94	79	30°30′
	Key West	108	57	90	79	24°30′
	Miami	214	47	90	79	25°50′
	Orlando	766	38	93	78	28°30′
	Pensacola	1,463	29	93	79	30°30′
	Tallahassee	1,485	30	92	78	30°20′
	Tampa	683	40	91	79	28°00′
	West Palm Beach	253	45	91	79	26°40′
GA	Athens	2,929	22	92	77	34°00′
	Atlanta	2,961	22	92	76	33°40′
	Augusta	2,397	23	95	79	33°20′
	Columbus	2,383	24	93	78	32°30′
	Macon	2,136	25	93	78	32°40′
	Rome	3,326	22	93	78	34°20′
	Savannah	1,819	27	93	79	32°10′
HI	Hilo	0	62	83	74	19°40′
	Honolulu	0	63	86	75	21°20′
ID	Boise	5,809	10	94	66	43°30′
	Lewiston	5,542	6	93	66	46°20′
	Pocatello	7,033	-1	91	63	43°00′
IL	Chicago (Midway)	6,155	0	91	75	41°50′
	Chicago (O'Hare)	6,639	-4	89	76	42°00′
	Chicago[d]	5,882	2	91	77	41°50′
	Moline	6,408	-4	91	77	41°30′
	Peoria	6,025	-4	89	76	40°40′
	Rockford	6,830	-4	89	76	42°10′
	Springfield	5,429	2	92	77	39°50′
IN	Evansville	4,435	9	93	78	38°00′
	Fort Wayne	6,205	1	89	75	41°00′
	Indianapolis	5,699	2	90	76	39°40′
	South Bend	6,439	1	89	75	41°40′
IA	Burlington	6,114	-3	91	77	40°50′
	Des Moines	6,588	-5	91	77	41°30′
	Dubuque	7,376	-7	88	75	42°20′
	Sioux City	6,951	-7	92	77	42°20′
	Waterloo	7,320	-10	89	77	42°30′
KS	Dodge City	4,986	5	97	73	37°50′
	Goodland	6,141	0	96	70	39°20′
	Topeka	5,182	4	96	78	39°00′
	Wichita	4,620	7	98	76	37°40′
KY	Covington	5,265	6	90	75	39°00′
	Lexington	4,683	8	91	76	38°00′
	Louisville	4,660	10	93	77	38°10′
LA	Alexandria	1,921	27	94	79	31°20′
	Baton Rouge	1,560	29	93	80	30°30′
	Lake Charles	1,459	31	93	79	30°10′
	New Orleans	1,385	33	92	80	30°00′
	Shreveport	2,184	25	96	79	32°30′

(continued)

APPENDIX D

TABLE D101—continued
DEGREE DAY AND DESIGN TEMPERATURES[a] FOR CITIES IN THE UNITED STATES

STATE	STATION[b]	HEATING DEGREE DAYS (yearly total)	DESIGN TEMPERATURES			DEGREES NORTH LATITUDE[c]
			Winter	Summer		
			97$\frac{1}{2}$%	Dry bulb 2$\frac{1}{2}$%	Wet bulb 2$\frac{1}{2}$%	
ME	Caribou	9,767	-13	81	69	46°50'
	Portland	7,511	-1	84	72	43°40'
MD	Baltimore	4,654	13	91	77	39°10'
	Baltimore[d]	4,111	17	89	78	39°20'
	Frederick	5,087	12	91	77	39°20'
MA	Boston	5,634	9	88	74	42°20'
	Pittsfield	7,578	-3	84	72	42°30'
	Worcester	6,969	4	84	72	42°20'
MI	Alpena	8,506	-6	85	72	45°00'
	Detroit (City)	6,232	6	88	74	42°20'
	Escanaba[d]	8,481	-7	83	71	45°40'
	Flint	7,377	1	87	74	43°00'
	Grand Rapids	6,894	5	88	74	42°50'
	Lansing	6,909	1	87	74	42°50'
	Marquette[d]	8,393	-8	81	70	46°30'
	Muskegon	6,696	6	84	73	43°10'
	Sault Ste. Marie	9,048	-8	81	70	46°30'
MN	Duluth	10,000	-16	82	70	46°50'
	Minneapolis	8,382	-12	89	5	44°50'
	Rochester	8,295	-12	87	75	44°00'
MS	Jackson	2,239	25	95	78	32°20'
	Meridian	2,289	23	95	79	32°20'
	Vicksburg[d]	2,041	26	95	80	32°20'
MO	Columbia	5,046	4	94	77	39°00'
	Kansas City	4,711	6	96	77	39°10'
	St. Joseph	5,484	2	93	79	39°50'
	St. Louis	4,900	6	94	77	38°50'
	St. Louis[d]	4,484	8	94	77	38°40'
	Springfield	4,900	9	93	77	37°10'
MT	Billings	7,049	-10	91	66	45°50'
	Great Falls	7,750	-15	88	62	47°30'
	Helena	8,129	-16	88	62	46°40'
	Missoula	8,125	-6	88	63	46°50'
NE	Grand Island	6,530	-3	94	74	41°00'
	Lincoln[d]	5,864	-2	95	77	40°50'
	Norfolk	6,979	-4	93	77	42°00'
	North Platte	6,684	-4	94	72	41°10'
	Omaha	6,612	-3	91	77	41°20'
	Scottsbluff	6,673	-3	92	68	41°50'
NV	Elko	7,433	-2	92	62	40°50'
	Ely	7,733	-4	87	59	39°10'
	Las Vegas	2,709	28	106	70	36°10'
	Reno	6,332	10	92	62	39°30'
	Winnemucca	6,761	3	94	62	40°50'
NH	Concord	7,383	-3	87	73	43°10'
NJ	Atlantic City	4,812	13	89	77	39°30'
	Newark	4,589	14	91	76	40°40'
	Trenton[d]	4,980	14	88	76	40°10'

(continued)

APPENDIX D

TABLE D101—continued
DEGREE DAY AND DESIGN TEMPERATURES[a] FOR CITIES IN THE UNITED STATES

STATE	STATION[b]	HEATING DEGREE DAYS (yearly total)	DESIGN TEMPERATURES			DEGREES NORTH LATITUDE[c]
			Winter	Summer		
			97½%	Dry bulb 2½%	Wet bulb 2½%	
NY	Albany	6,875	-1	88	74	42°50'
	Albany[d]	6,201	1	88	74	42°50'
	Binghamton	7,286	1	83	72	42°10'
	Buffalo	7,062	6	85	73	43°00'
	NY (Central Park)[d]	4,871	15	89	75	40°50'
	NY (Kennedy)	5,219	15	87	75	40°40'
	NY (LaGuardia)	4,811	15	89	75	40°50'
	Rochester	6,748	5	88	73	43°10'
	Schenectady[d]	6,650	1	87	74	42°50'
	Syracuse	6,756	2	87	73	43°10'
NC	Charlotte	3,181	22	93	76	35°10'
	Greensboro	3,805	18	91	76	36°10'
	Raleigh	3,393	20	92	77	35°50'
	Winston-Salem	3,595	20	91	75	36°10'
ND	Bismarck	8,851	-19	91	71	46°50'
	Devils Lake[d]	9,901	-21	88	71	48°10'
	Fargo	9,226	-18	89	74	46°50'
	Williston	9,243	-21	88	70	48°10'
OH	Akron-Canton	6,037	6	86	73	41°00'
	Cincinnati[d]	4,410	6	90	75	39°10'
	Cleveland	6,351	5	88	74	41°20'
	Columbus	5,660	5	90	75	40°00'
	Dayton	5,622	4	89	75	39°50'
	Mansfield	6,403	5	87	74	40°50'
	Sandusky[d]	5,796	6	91	74	41°30'
	Toledo	6,494	1	88	75	41°40'
	Youngstown	6,417	4	86	73	41°20'
OK	Oklahoma City	3,725	13	97	77	35°20'
	Tulsa	3,860	13	98	78	36°10'
OR	Eugene	4,726	22	89	67	44°10'
	Medford	5,008	23	94	68	42°20'
	Portland	4,635	23	85	67	45°40'
	Portland[d]	4,109	24	86	67	45°30'
	Salem	4,754	23	88	68	45°00'
PA	Allentown	5,810	9	88	75	40°40'
	Erie	6,451	9	85	74	42°10'
	Harrisburg	5,251	11	91	76	40°10'
	Philadelphia	5,144	14	90	76	39°50'
	Pittsburgh	5,987	5	86	73	40°30'
	Pittsburgh[d]	5,053	7	88	73	40°30'
	Reading[d]	4,945	13	89	75	40°20'
	Scranton	6,254	5	87	73	41°20'
	Williamsport	5,934	7	89	74	41°10'
RI	Providence	5,954	9	86	74	41°40'
SC	Charleston	2,033	27	91	80	32°50'
	Charleston[d]	1,794	28	92	80	32°50'
	Columbia	2,484	24	95	78	34°00'

(continued)

TABLE D101—continued
DEGREE DAY AND DESIGN TEMPERATURES[a] FOR CITIES IN THE UNITED STATES

STATE	STATION[b]	HEATING DEGREE DAYS (yearly total)	DESIGN TEMPERATURES			DEGREES NORTH LATITUDE[c]
			Winter	Summer		
			97 1/2%	Dry bulb 2 1/2%	Wet bulb 2 1/2%	
NM	Albuquerque	4,348	16	94	65	35°00'
	Raton	6,228	1	89	64	36°50'
	Roswell	3,793	18	98	70	33°20'
	Silver City	3,705	10	94	64	32°40'
SD	Huron	8,223	-14	93	75	44°30'
	Rapid City	7,345	-7	92	69	44°00'
	Sioux Falls	7,839	-11	91	75	43°40'
TN	Bristol	4,143	14	89	75	36°30'
	Chattanooga	3,254	18	93	77	35°00'
	Knoxville	3,494	19	92	76	35°50'
	Memphis	3,232	18	95	79	35°00'
	Nashville	3,578	14	94	77	36°10'
TX	Abilene	2,624	20	99	74	32°30'
	Austin	1,711	28	98	77	30°20'
	Dallas	2,363	22	100	78	32°50'
	El Paso	2,700	24	98	68	31°50'
	Houston	1,396	32	94	79	29°40'
	Midland	2,591	21	98	72	32°00'
	San Angelo	2,255	22	99	74	31°20'
	San Antonio	1,546	30	97	76	29°30'
	Waco	2,030	26	99	78	31°40'
	Wichita Falls	2,832	18	101	76	34°00'
UT	Salt Lake City	6,052	8	95	65	40°50'
VT	Burlington	8,269	-7	85	72	44°30'
VA	Lynchburg	4,166	16	90	76	37°20'
	Norfolk	3,421	22	91	78	36°50'
	Richmond	3,865	17	92	78	37°30'
	Roanoke	4,150	16	91	74	37°20'
WA	Olympia	5,236	22	83	66	47°00'
	Seattle-Tacoma	5,145	26	80	64	47°30'
	Seattle[d]	4,424	27	82	67	47°40'
	Spokane	6,655	2	90	64	47°40'
WV	Charleston	4,476	11	90	75	38°20'
	Elkins	5,675	6	84	72	38°50'
	Huntington	4,446	10	91	77	38°20'
	Parkersburg[d]	4,754	11	90	76	39°20'
WI	Green Bay	8,029	-9	85	74	44°30'
	La Crosse	7,589	-9	88	75	43°50'
	Madison	7,863	-7	88	75	43°10'
	Milwaukee	7,635	-4	87	74	43°00'
WY	Casper	7,410	-5	90	61	42°50'
	Cheyenne	7,381	-1	86	62	41°10'
	Lander	7,870	-11	88	63	42°50'
	Sheridan	7,680	-8	91	65	44°50'

a. All data were extracted from the 1985 ASHRAE Handbook, Fundamentals Volume.
b. Design data developed from airport temperature observations unless noted.
c. Latitude is given to the nearest 10 minutes. For example, the latitude for Miami, Florida, is given as 25°50', or 25 degrees 50 minutes.
d. Design data developed from office locations within an urban area, not from airport temperature observations.

APPENDIX E

SIZING OF WATER PIPING SYSTEM

The provisions contained in this appendix are not mandatory unless specifically referenced in the adopting ordinance.

User note:

About this appendix: *The sizing of water service and water distribution piping is not specified in Chapter 6 as it is left up to the designer of the system with the code official approving the design method. Appendix E provides several methods that could be used by a system designer.*

SECTION E101
GENERAL

E101.1 Scope.

E101.1.1 This appendix outlines two procedures for sizing a water piping system (see Sections E103.3 and E201.1). The design procedures are based on the minimum static pressure available from the supply source, the head changes in the system caused by friction and elevation, and the rates of flow necessary for operation of various fixtures.

E101.1.2 Because of the variable conditions encountered in hydraulic design, it is impractical to specify definite and detailed rules for sizing of the water piping system. Accordingly, other sizing or design methods conforming to good engineering practice standards are acceptable alternatives to those presented herein.

SECTION E102
INFORMATION REQUIRED

E102.1 Preliminary. Obtain the necessary information regarding the minimum daily static service pressure in the area where the building is to be located. If the building supply is to be metered, obtain information regarding friction loss relative to the rate of flow for meters in the range of sizes likely to be used. Friction loss data can be obtained from most manufacturers of water meters.

E102.2 Demand load.

E102.2.1 Estimate the supply demand of the building main and the principal branches and risers of the system by totaling the corresponding demand from the applicable part of Table E103.3(3).

E102.2.2 Estimate continuous supply demands in gallons per minute (L/m) for items such as lawn sprinklers and air conditioners, and add the sum to the total demand for fixtures. The result is the estimated supply demand for the building supply.

SECTION E103
SELECTION OF PIPE SIZE

E103.1 General. Decide from Table 604.3 what is the desirable minimum residual pressure that should be maintained at the highest fixture in the supply system. If the highest group of fixtures contains flushometer valves, the pressure for the group should be not less than 15 pounds per square inch (psi) (103.4 kPa) flowing. For flush tank supplies, the available pressure should be not less than 8 psi (55.2 kPa) flowing, except blowout action fixtures must be not less than 25 psi (172.4 kPa) flowing.

E103.2 Pipe sizing.

E103.2.1 Pipe sizes can be selected according to the following procedure or by other design methods conforming to acceptable engineering practice and *approved* by the administrative authority. The sizes selected must not be less than the minimum required by this code.

E103.2.2 Water pipe sizing procedures are based on a system of pressure requirements and losses, the sum of which must not exceed the minimum pressure available at the supply source. These pressures are as follows:

1. Pressure required at fixture to produce required flow. See Sections 604.3 and 604.5.

2. Static pressure loss or gain (due to head) is computed at 0.433 psi per foot (9.8 kPa/m) of elevation change.

    Example: Assume that the highest fixture supply outlet is 20 feet (6096 mm) above or below the supply source. This produces a static pressure differential of 20 feet by 0.433 psi/foot (2096 mm by 9.8 kPa/m) and an 8.66 psi (59.8 kPa) loss.

3. Loss through water meter. The friction or pressure loss can be obtained from meter manufacturers.

4. Loss through taps in water main.

5. Losses through special devices such as filters, softeners, backflow prevention devices and pressure regulators. These values must be obtained from the manufacturers.

6. Loss through valves and fittings. Losses for these items are calculated by converting to equivalent length of piping and adding to the total pipe length.

7. Loss due to pipe friction can be calculated where the pipe size, the pipe length and the flow through the pipe are known. With these three items, the friction loss can be determined. For piping flow charts not included, use manufacturers' tables and velocity recommendations.

    Note: For the purposes of all examples, the following metric conversions are applicable:

    1 cubic foot per minute = 0.4719 L/s

APPENDIX E

1 square foot = 0.0929 m²
1 degree = 0.0175 rad
1 pound per square inch = 6.895 kPa
1 inch = 25.4 mm
1 foot = 304.8 mm
1 gallon per minute = 3.785 L/m

E103.3 Segmented loss method. The size of water service mains, *branch* mains and risers by the segmented loss method must be determined according to water supply demand [gpm (L/m)], available water pressure [psi (kPa)] and friction loss caused by the water meter and *developed length* of pipe [feet (m)], including equivalent length of fittings. This design procedure is based on the following parameters:

- Calculates the friction loss through each length of the pipe.
- Based on a system of pressure losses, the sum of which must not exceed the minimum pressure available at the street main or other source of supply.
- Pipe sizing based on estimated peak demand, total pressure losses caused by difference in elevation, equipment, *developed length* and pressure required at most remote fixture, loss through taps in water main, losses through fittings, filters, backflow prevention devices, valves and pipe friction.

Because of the variable conditions encountered in hydraulic design, it is impractical to specify definite and detailed rules for sizing of the water piping system. Current sizing methods do not address the differences in the probability of use and flow characteristics of fixtures between types of occupancies. Creating an exact model of predicting the demand for a building is impossible and final studies assessing the impact of water conservation on demand are not yet complete. The following steps are necessary for the segmented loss method.

1. **Preliminary.** Obtain the necessary information regarding the minimum daily static service pressure in the area where the building is to be located. If the building supply is to be metered, obtain information regarding friction loss relative to the rate of flow for meters in the range of sizes to be used. Friction loss data can be obtained from manufacturers of water meters. It is essential that enough pressure be available to overcome all system losses caused by friction and elevation so that plumbing fixtures operate properly. Section 604.6 requires the water distribution system to be designed for the minimum pressure available taking into consideration pressure fluctuations. The lowest pressure must be selected to guarantee a continuous, adequate supply of water. The lowest pressure in the public main usually occurs in the summer because of lawn sprinkling and supplying water for air-conditioning cooling towers. Future demands placed on the public main as a result of large growth or expansion should be considered. The available pressure will decrease as additional loads are placed on the public system.

2. **Demand load.** Estimate the supply demand of the building main and the principal branches and risers of the system by totaling the corresponding demand from the applicable part of Table E103.3(3). When estimating peak demand sizing methods typically use water supply fixture units (w.s.f.u.) [see Table E103.3(2)]. This numerical factor measures the load-producing effect of a single plumbing fixture of a given kind. The use of such fixture units can be applied to a single basic probability curve (or table), found in the various sizing methods [Table E103.3(3)]. The fixture units are then converted into gallons per minute (L/m) flow rate for estimating demand.

    2.1. Estimate continuous supply demand in gallons per minute (L/m) for items such as lawn sprinklers and air conditioners, and add the sum to the total demand for fixtures. The result is the estimated supply demand for the building supply. Fixture units cannot be applied to constant use fixtures such as hose bibbs, lawn sprinklers and air conditioners. These types of fixtures must be assigned the gallon per minute (L/m) value.

3. **Selection of pipe size.** This water pipe sizing procedure is based on a system of pressure requirements and losses, the sum of which must not exceed the minimum pressure available at the supply source. These pressures are as follows:

    3.1. Pressure required at the fixture to produce required flow. See Sections 604.3 and 604.5.

    3.2. Static pressure loss or gain (because of head) is computed at 0.433 psi per foot (9.8 kPa/m) of elevation change.

    3.3. Loss through a water meter. The friction or pressure loss can be obtained from the manufacturer.

    3.4. Loss through taps in water main [see Table E103.3(4)].

    3.5. Losses through special devices such as filters, softeners, backflow prevention devices and pressure regulators. These values must be obtained from the manufacturers.

    3.6. Loss through valves and fittings [see Tables E103.3(5) and E103.3(6)]. Losses for these items are calculated by converting to equivalent length of piping and adding to the total pipe length.

    3.7. Loss caused by pipe friction can be calculated where the pipe size, the pipe length and the flow through the pipe are known. With these three items, the friction loss can be determined using Figures E103.3(2) through E103.3(7). When using charts, use pipe inside diameters. For piping flow charts not included, use manufacturers' tables and velocity recommendations. Before attempting to size any water supply system, it is necessary to gather preliminary information that includes available pressure, piping material, select design velocity, elevation differences and *developed length* to most remote fixture. The water supply system is divided into sections at major changes in elevation or where *branches* lead to fixture groups. The peak demand must be determined in each part of the hot and cold water supply system that includes the corresponding water supply fixture unit and conversion to gallons per minute (L/m) flow rate to be expected through each section. Sizing methods require the determination of the "most hydraulically remote" fixture to compute the pressure loss caused by pipe and fittings. The hydraulically remote fixture represents the most downstream fixture along the circuit of piping requiring the most available pressure to operate properly. Consideration must be given to all pressure demands and losses, such as friction caused by pipe, fittings and equipment, elevation and the residual pressure required by Table 604.3. The two most common and frequent complaints about the water supply system operation are lack of adequate pressure and noise.

Problem: What size Type L copper water pipe, service and distribution will be required to serve a two-story factory building having on each floor, back-to-back, two toilet rooms each equipped with hot and cold water? The highest fixture is 21 feet (6401 mm) above the street main, which is tapped with a 2-inch (51 mm) corporation cock at which point the minimum pressure is 55 psi (379.2 kPa). In the building basement, a 2-inch (51 mm) meter with a maximum pressure drop of 11 psi (75.8 kPa) and 3-inch (76 mm) reduced pressure principle backflow preventer with a maximum pressure drop of 9 psi (621 kPa) are to be installed. The system is shown by Figure E103.3(1). To be determined are the pipe sizes for the service main and the cold and hot water distribution pipes.

Solution: A tabular arrangement such as shown in Table E103.3(1) should first be constructed. The steps to be followed are indicated by the tabular arrangement itself as they are in sequence, Columns 1 through 10 and Lines A through L.

Step 1

Columns 1 and 2: Divide the system into sections breaking at major changes in elevation or where *branches* lead to fixture groups. After point B [see Figure E103.3(1)], separate consideration will be given to the hot and cold water piping. Enter the sections to be considered in the service and cold water piping in Column 1 of the tabular arrangement. Column 1 of Table E103.3(1) provides a line-by-line recommended tabular arrangement for use in solving pipe sizing.

The objective in designing the water supply system is to ensure an adequate water supply and pressure to all fixtures and equipment. Column 2 provides the pounds per square inch (psi) to be considered separately from the minimum pressure available at the main. Losses to take into consideration are the following: the differences in elevation between the water supply source and the highest water supply outlet, meter pressure losses, the tap in main loss, special fixture devices such as water softeners and backflow prevention devices and the pressure required at the most remote fixture outlet. The difference in elevation can result in an increase or decrease in available pressure at the main. Where the water supply outlet is located above the source, this results in a loss in the available pressure and is subtracted from the pressure at the water source. Where the highest water supply outlet is located below the water supply source, there will be an increase in pressure that is added to the available pressure of the water source.

Column 3: According to Table E103.3(3), determine the gpm (L/m) of flow to be expected in each section of the system. These flows range from 28.6 to 108 gpm. Load values for fixtures must be determined as water supply fixture units and then converted to a gallon-per-minute (gpm) rating to determine peak demand. When calculating peak demands, the water supply fixture units are added and then converted to the gallon-per-minute rating. For continuous flow fixtures such as hose bibbs and lawn sprinkler systems, add the gallon-per-minute demand to the intermittent demand of fixtures. For example, a total of 120 water supply fixture units is converted to a demand of 48 gallons per minute. Two hose bibbs × 5 gpm demand = 10 gpm. Total gpm rating = 48.0 gpm + 10 gpm = 58.0 gpm demand.

Step 2

Line A: Enter the minimum pressure available at the main source of supply in Column 2. This is 55 psi (379.2 kPa). The local water authorities generally keep records of pressures at different times of day and year. The available pressure can be checked from nearby buildings or from fire department hydrant checks.

Line B: Determine from Table 604.3 the highest pressure required for the fixtures on the system, which is 15 psi (103.4 kPa), to operate a flushometer valve. The most remote fixture outlet is necessary to compute the pressure loss caused by pipe and fittings, and represents the most downstream fixture along the circuit of piping requiring the available pressure to operate properly as indicated by Table 604.3.

Line C: Determine the pressure loss for the meter size given or assumed. The total water flow from the main through the service as determined in Step 1 will serve to aid in the meter selected. There are three common types of water meters; the pressure losses are determined by the American Water Works Association Standards for displacement type, compound type and turbine type. The maximum pressure loss of such devices takes into consideration the meter size, safe operating capacity (gpm) and maximum rates for continuous operations (gpm). Typically, equipment imparts greater pressure losses than piping.

Line D: Select from Table E103.3(4) and enter the pressure loss for the tap size given or assumed. The loss of pressure through taps and tees in pounds per square inch (psi) is based on the total gallon-per-minute flow rate and size of the tap.

APPENDIX E

Line E: Determine the difference in elevation between the main and source of supply and the highest fixture on the system. Multiply this figure, expressed in feet, by 0.43 psi (2.9 kPa). Enter the resulting psi loss on Line E. The difference in elevation between the water supply source and the highest water supply outlet has a significant impact on the sizing of the water supply system. The difference in elevation usually results in a loss in the available pressure because the water supply outlet is generally located above the water supply source. The loss is caused by the pressure required to lift the water to the outlet. The pressure loss is subtracted from the pressure at the water source. Where the highest water supply outlet is located below the water source, there will be an increase in pressure that is added to the available pressure of the water source.

Lines F, G and H: The pressure losses through filters, backflow prevention devices or other special fixtures must be obtained from the manufacturer or estimated and entered on these lines. Equipment such as backflow prevention devices, check valves, water softeners, instantaneous or tankless water heaters, filters and strainers can impart a much greater pressure loss than the piping. The pressure losses can range from 8 psi to 30 psi.

Step 3
Line I: The sum of the pressure requirements and losses that affect the overall system (Lines B through H) is entered on this line. Summarizing the steps, all of the system losses are subtracted from the minimum water pressure. The remainder is the pressure available for friction, defined as the energy available to push the water through the pipes to each fixture. This force can be used as an average pressure loss, as long as the pressure available for friction is not exceeded. Saving a certain amount for available water supply pressures as an area incurs growth, or because of aging of the pipe or equipment added to the system is recommended.

Step 4
Line J: Subtract Line I from Line A. This gives the pressure that remains available from overcoming friction losses in the system. This figure is a guide to the pipe size that is chosen for each section, incorporating the total friction losses to the most remote outlet (measured length is called *developed length*).

> **Exception:** Where the main is above the highest fixture, the resulting psi must be considered a pressure gain (static head gain) and omitted from the sums of Lines B through H and added to Line J.

The maximum friction head loss that can be tolerated in the system during peak demand is the difference between the static pressure at the highest and most remote outlet at no-flow conditions and the minimum flow pressure required at that outlet. If the losses are within the required limits, then every run of pipe will be within the required friction head loss. Static pressure loss is the most remote outlet in feet × 0.433 = loss in psi caused by elevation differences.

Step 5
Column 4: Enter the length of each section from the main to the most remote outlet (at Point E). Divide the water supply system into sections breaking at major changes in elevation or where *branches* lead to fixture groups.

Step 6
E103.3.3. Selection of pipe size, Step 6 Column 5: When selecting a trial pipe size, the length from the water service or meter to the most remote fixture outlet must be measured to determine the *developed length*. However, in systems having a flushometer valve or temperature controlled shower at the topmost floors the *developed length* would be from the water meter to the most remote flushometer valve on the system. A rule of thumb is that size will become progressively smaller as the system extends farther from the main source of supply. The following formula is an acceptable method to determine trial pipe size:

Line J: (Pressure available to overcome pipe friction) × 100/equivalent length of run total *developed length* to most remote fixture × percentage factor of 1.5 (note: a percentage factor is used only as an estimate for friction losses imposed for fittings for initial trial pipe size) = psi (average pressure drops per 100 feet of pipe).

For trial pipe size, see Figure E 103.3(3) (Type L copper) based on 2.77 psi and a 108 gpm = $2^1/_2$ inches. To determine the equivalent length of run to the most remote outlet, the *developed length* is determined and added to the friction losses for fittings and valves. The *developed lengths* of the designated pipe sections are as follows:

A - B	54 ft
B - C	8 ft
C - D	13 ft
D - E	150 ft
Total developed length = 225 ft	

The equivalent length of the friction loss in fittings and valves must be added to the *developed length* (most remote outlet). Where the size of fittings and valves is not known, the added friction loss should be approximated. A general rule that has been used is to add 50 percent of the *developed length* to allow for fittings and valves. For example, the equivalent length of run equals the *developed length* of run (225 ft × 1.5 = 338 ft). The total equivalent length of run for determining a trial pipe size is 338 feet.

> **Example:** 9.36 (pressure available to overcome pipe friction) × 100/338 (equivalent length of run = 225 × 1.5) = 2.77 psi (average pressure drop per 100 feet of pipe).

Step 7
Column 6: Select from Table E103.3(6) the equivalent lengths for the trial pipe size of fittings and valves on each pipe section. Enter the sum for each section in Column 6. (The number of fittings to be used in this example must be an estimate.) The equivalent length of piping is the *developed length* plus the equivalent lengths of pipe corre-

TABLE E.1

COLD WATER PIPE SECTION	FITTINGS/VALVES	PRESSURE LOSS EXPRESSED AS EQUIVALENT LENGTH OF TUBE (feet)	HOT WATER PIPE SECTION	FITTINGS/ VALVES	PRESSURE LOSS EXPRESSED AS EQUIVALENT OF TUBE (feet)
A-B	3-2$\frac{1}{2}$" Gate valves	3	A-B	3-2$\frac{1}{2}$" Gate valves	3
	1-2$\frac{1}{2}$" Side branch tee	12		1-2$\frac{1}{2}$" Side branch tee	12
B-C	1-2$\frac{1}{2}$" Straight run tee	0.5	B-C	1-2" Straight run tee	7
				1-2" 90-degree ell	0.5
C-F	1-2$\frac{1}{2}$" Side branch tee	12	C-F	1-1$\frac{1}{2}$" Side branch tee	7
C-D	1-2$\frac{1}{2}$" 90-degree ell	7	C-D	1-1$\frac{1}{2}$" 90-degree ell	4
D-E	1-2$\frac{1}{2}$" Side branch tee	12	D-E	1-1$\frac{1}{2}$" Side branch tee	7

For SI: 1 foot = 304.8 mm, 1 inch = 25.4 mm.

sponding to friction head losses for fittings and valves. Where the size of fittings and valves is not known, the added friction head losses must be approximated. An estimate for this example is found in Table E.1.

Step 8

Column 7: Add the figures from Column 4 and Column 6, and enter in Column 7. Express the sum in hundreds of feet.

Step 9

Column 8: Select from Figure E103.3(3) the friction loss per 100 feet (30 480 mm) of pipe for the gallon-per-minute flow in a section (Column 3) and trial pipe size (Column 5). Maximum friction head loss per 100 feet is determined on the basis of total pressure available for friction head loss and the longest equivalent length of run. The selection is based on the gallon-per-minute demand, the uniform friction head loss and the maximum design velocity. Where the size indicated by hydraulic table indicates a velocity in excess of the selected velocity, a size must be selected that produces the required velocity.

Step 10

Column 9: Multiply the figures in Columns 7 and 8 for each section and enter in Column 9.

Total friction loss is determined by multiplying the friction loss per 100 feet (30 480 mm) for each pipe section in the total *developed length* by the pressure loss in fittings expressed as equivalent length in feet. Note: Section C-F should be considered in the total pipe friction losses only if greater loss occurs in Section C-F than in pipe section D-E. Section C-F is not considered in the total *developed length*. Total friction loss in equivalent length is determined in Table E.2.

Step 11

Line K: Enter the sum of the values in Column 9. The value is the total friction loss in equivalent length for each designated pipe section.

Step 12

Line L: Subtract Line J from Line K and enter in Column 10.

The result should always be a positive or plus figure. If it is not, repeat the operation using Columns 5, 6, 8 and 9 until a balance or near balance is obtained. If the difference between Lines J and K is a high positive number, it is an indication that the pipe sizes are too large and should be reduced, thus saving materials. In such a case, the operations using Columns 5, 6, 8 and 9 should again be repeated.

The total friction losses are determined and subtracted from the pressure available to overcome pipe friction for trial pipe size. This number is critical as it provides a guide to whether the pipe size selected is too large and the process should be repeated to obtain an economically designed system.

Answer: The final figures entered in Column 5 become the design pipe size for the respective sections. Repeating this operation a second time using the same sketch but considering the demand for hot water, it is possible to size the hot water distribution piping. This has been worked up as a part of the overall problem in the tabular arrangement used for sizing the service and water distribution piping. Note that consideration must be given to the pressure losses from the street main to the water heater (Section A-B) in determining the hot water pipe sizes.

TABLE E.2

PIPE SECTIONS	FRICTION LOSS EQUIVALENT LENGTH (feet)	
	Cold Water	Hot Water
A-B	0.69 × 3.2 = 2.21	0.69 × 3.2 = 2.21
B-C	0.085 × 3.1 = 0.26	0.16 × 1.4 = 0.22
C-D	0.20 × 1.9 = 0.38	0.17 × 3.2 = 0.54
D-E	1.62 × 1.9 = 3.08	1.57 × 3.2 = 5.02
Total pipe friction losses (Line K)	5.93	7.99

For SI: 1 foot = 304.8 mm, 1 gpm = 3.785 L/m.

APPENDIX E

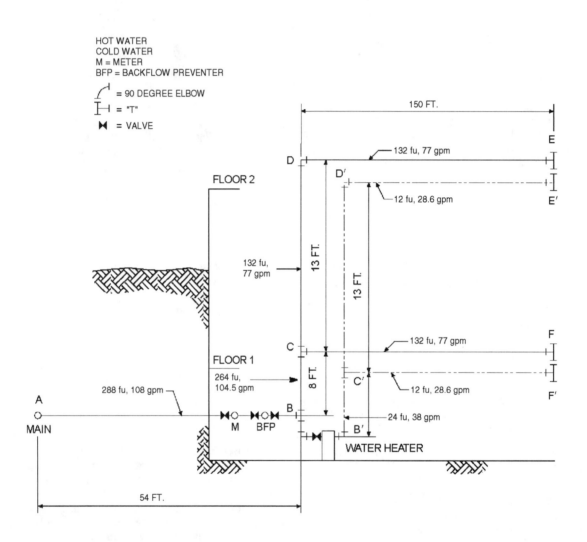

For SI: 1 foot = 304.8 mm, 1 gpm = 3.785 L/m.

**FIGURE E103.3(1)
EXAMPLE-SIZING**

APPENDIX E

TABLE E103.3(1)
RECOMMENDED TABULAR ARRANGEMENT FOR USE IN SOLVING PIPE SIZING PROBLEMS

COLUMN	1		2	3	4	5	6	7	8	9	10
Line	Description		Lb per square inch (psi)	Gal. per min through section	Length of section (feet)	Trial pipe size (inches)	Equivalent length of fittings and valves (feet)	Total equivalent length col. 4 and col. 6 (100 feet)	Friction loss per 100 feet of trial size pipe (psi)	Friction loss in equivalent length col. 8 x col. 7 (psi)	Excess pressure over friction losses (psi)
A	Service and cold water distribution piping[a]	Minimum pressure available at main........ 55.00									
B		Highest pressure required at a fixture (Table 604.3)........................ 15.00									
C		Meter loss 2″ meter 11.00									
D		Tap in main loss 2″ tap [Table E103.3(4)].................... 1.61									
E		Static head loss 21 × 43 psi 9.03									
F		Special fixture loss backflow preventer........................ 9.00									
G		Special fixture loss—Filter 0.00									
H		Special fixture loss—Other 0.00									
I		Total overall losses and requirements (Sum of Lines B through H)............. 45.64									
J		Pressure available to overcome pipe friction (Line A minus Lines B to H)............. 9.36									
	DESIGNATION Pipe section (from diagram) Cold water Distribution piping	FU....................264									
		AB....................288	108.0	54	2½	15.00	0.69	3.2	2.21	—
		BC....................264	104.5	8	2½	0.5	0.85	3.1	0.26	—
		CD....................132	77.0	13	2½	7.00	0.20	1.9	0.38	—
		CF[b]..................132	77.0	150	2½	12.00	1.62	1.9	3.08	—
		DE[b]..................132	77.0	150	2½	12.00	1.62	1.9	3.08	—
K	Total pipe friction losses (cold)		—	—	—	—	—	—	—	5.93	—
L	Difference (Line J minus Line K)		—	—	—	—	—	—	—	—	3.43
	Pipe section (from diagram) Diagram Hot water Distribution Piping	A'B'...................288	108.0	54	2½	12.00	0.69	3.3	2.21	—
		B'C'....................24	38.0	8	2	7.5	0.16	1.4	0.22	—
		C'D'....................12	28.6	13	1½	4.0	0.17	3.2	0.54	—
		C'F'[b]..................12	28.6	150	1½	7.00	1.57	3.2	5.02	—
		D'E'[b]..................12	28.6	150	1½	7.00	1.57	3.2	5.02	—
K	Total pipe friction losses (hot)		—	—	—	—	—	—	—	7.99	—
L	Difference (Line J minus Line K)		—	—	—	—	—	—	—	—	1.37

For SI: 1 inch = 25.4 mm, 1 foot = 304.8 mm, 1 psi = 6.895 kPa, 1 gpm = 3.785 L/m.

a. To be considered as pressure gain for fixtures below main (to consider separately, omit from "I" and add to "J").

b. To consider separately, in K use C-F only if greater loss than above.

APPENDIX E

TABLE E103.3(2)
LOAD VALUES ASSIGNED TO FIXTURES[a]

FIXTURE	OCCUPANCY	TYPE OF SUPPLY CONTROL	LOAD VALUES, IN WATER SUPPLY FIXTURE UNITS (wsfu)		
			Cold	Hot	Total
Bathroom group	Private	Flush tank	2.7	1.5	3.6
Bathroom group	Private	Flushometer valve	6.0	3.0	8.0
Bathtub	Private	Faucet	1.0	1.0	1.4
Bathtub	Public	Faucet	3.0	3.0	4.0
Bidet	Private	Faucet	1.5	1.5	2.0
Combination fixture	Private	Faucet	2.25	2.25	3.0
Dishwashing machine	Private	Automatic	—	1.4	1.4
Drinking fountain	Offices, etc.	$^3/_8$" valve	0.25	—	0.25
Kitchen sink	Private	Faucet	1.0	1.0	1.4
Kitchen sink	Hotel, restaurant	Faucet	3.0	3.0	4.0
Laundry trays (1 to 3)	Private	Faucet	1.0	1.0	1.4
Lavatory	Private	Faucet	0.5	0.5	0.7
Lavatory	Public	Faucet	1.5	1.5	2.0
Service sink	Offices, etc.	Faucet	2.25	2.25	3.0
Shower head	Public	Mixing valve	3.0	3.0	4.0
Shower head	Private	Mixing valve	1.0	1.0	1.4
Urinal	Public	1" flushometer valve	10.0	—	10.0
Urinal	Public	$^3/_4$" flushometer valve	5.0	—	5.0
Urinal	Public	Flush tank	3.0	—	3.0
Washing machine (8 lb)	Private	Automatic	1.0	1.0	1.4
Washing machine (8 lb)	Public	Automatic	2.25	2.25	3.0
Washing machine (15 lb)	Public	Automatic	3.0	3.0	4.0
Water closet	Private	Flushometer valve	6.0	—	6.0
Water closet	Private	Flush tank	2.2	—	2.2
Water closet	Public	Flushometer valve	10.0	—	10.0
Water closet	Public	Flush tank	5.0	—	5.0
Water closet	Public or private	Flushometer tank	2.0	—	2.0

For SI: 1 inch = 25.4 mm, 1 pound = 0.454 kg.

a. For fixtures not listed, loads should be assumed by comparing the fixture to one listed using water in similar quantities and at similar rates. The assigned loads for fixtures with both hot and cold water supplies are given for separate hot and cold water loads and for total load. The separate hot and cold water loads being three-fourths of the total load for the fixture in each case.

APPENDIX E

TABLE E103.3(3)
TABLE FOR ESTIMATING DEMAND

SUPPLY SYSTEMS PREDOMINANTLY FOR FLUSH TANKS			SUPPLY SYSTEMS PREDOMINANTLY FOR FLUSHOMETER VALVES		
Load	Demand		Load	Demand	
(Water supply fixture units)	(Gallons per minute)	(Cubic feet per minute)	(Water supply fixture units)	(Gallons per minute)	(Cubic feet per minute)
1	3.0	0.04104	—	—	—
2	5.0	0.0684	—	—	—
3	6.5	0.86892	—	—	—
4	8.0	1.06944	—	—	—
5	9.4	1.256592	5	15.0	2.0052
6	10.7	1.430376	6	17.4	2.326032
7	11.8	1.577424	7	19.8	2.646364
8	12.8	1.711104	8	22.2	2.967696
9	13.7	1.831416	9	24.6	3.288528
10	14.6	1.951728	10	27.0	3.60936
11	15.4	2.058672	11	27.8	3.716304
12	16.0	2.13888	12	28.6	3.823248
13	16.5	2.20572	13	29.4	3.930192
14	17.0	2.27256	14	30.2	4.037136
15	17.5	2.3394	15	31.0	4.14408
16	18.0	2.90624	16	31.8	4.241024
17	18.4	2.459712	17	32.6	4.357968
18	18.8	2.513184	18	33.4	4.464912
19	19.2	2.566656	19	34.2	4.571856
20	19.6	2.620128	20	35.0	4.6788
25	21.5	2.87412	25	38.0	5.07984
30	23.3	3.114744	30	42.0	5.61356
35	24.9	3.328632	35	44.0	5.88192
40	26.3	3.515784	40	46.0	6.14928
45	27.7	3.702936	45	48.0	6.41664
50	29.1	3.890088	50	50.0	6.684
60	32.0	4.27776	60	54.0	7.21872
70	35.0	4.6788	70	58.0	7.75344
80	38.0	5.07984	80	61.2	8.181216
90	41.0	5.48088	90	64.3	8.595624
100	43.5	5.81508	100	67.5	9.0234
120	48.0	6.41664	120	73.0	9.75864
140	52.5	7.0182	140	77.0	10.29336
160	57.0	7.61976	160	81.0	10.82808
180	61.0	8.15448	180	85.5	11.42964
200	65.0	8.6892	200	90.0	12.0312
225	70.0	9.3576	225	95.5	12.76644
250	75.0	10.026	250	101.0	13.50168

(continued)

APPENDIX E

TABLE E103.3(3)-continued
TABLE FOR ESTIMATING DEMAND

SUPPLY SYSTEMS PREDOMINANTLY FOR FLUSH TANKS			SUPPLY SYSTEMS PREDOMINANTLY FOR FLUSHOMETER VALVES		
Load	Demand		Load	Demand	
(Water supply fixture units)	(Gallons per minute)	(Cubic feet per minute)	(Water supply fixture units)	(Gallons per minute)	(Cubic feet per minute)
275	80.0	10.6944	275	104.5	13.96956
300	85.0	11.3628	300	108.0	14.43744
400	105.0	14.0364	400	127.0	16.97736
500	124.0	16.57632	500	143.0	19.11624
750	170.0	22.7256	750	177.0	23.66136
1,000	208.0	27.80544	1,000	208.0	27.80544
1,250	239.0	31.94952	1,250	239.0	31.94952
1,500	269.0	35.95992	1,500	269.0	35.95992
1,750	297.0	39.70296	1,750	297.0	39.70296
2,000	325.0	43.446	2,000	325.0	43.446
2,500	380.0	50.7984	2,500	380.0	50.7984
3,000	433.0	57.88344	3,000	433.0	57.88344
4,000	525.0	70.182	4,000	525.0	70.182
5,000	593.0	79.27224	5,000	593.0	79.27224

For SI: 1 inch = 25.4 mm, 1 gallon per minute = 3.785 L/m, 1 cubic foot per minute = 0.28 m^3 per minute.

TABLE E103.3(4)
LOSS OF PRESSURE THROUGH TAPS AND TEES IN POUNDS PER SQUARE INCH (psi)

GALLONS PER MINUTE	SIZE OF TAP OR TEE (inches)						
	$5/8$	$3/4$	1	$1^1/_4$	$1^1/_2$	2	3
10	1.35	0.64	0.18	0.08	—	—	—
20	5.38	2.54	0.77	0.31	0.14	—	—
30	12.10	5.72	1.62	0.69	0.33	0.10	—
40	—	10.20	3.07	1.23	0.58	0.18	—
50	—	15.90	4.49	1.92	0.91	0.28	—
60	—	—	6.46	2.76	1.31	0.40	—
70	—	—	8.79	3.76	1.78	0.55	0.10
80	—	—	11.50	4.90	2.32	0.72	0.13
90	—	—	14.50	6.21	2.94	0.91	0.16
100	—	—	17.94	7.67	3.63	1.12	0.21
120	—	—	25.80	11.00	5.23	1.61	0.30
140	—	—	35.20	15.00	7.12	2.20	0.41
150	—	—	—	17.20	8.16	2.52	0.47
160	—	—	—	19.60	9.30	2.92	0.54
180	—	—	—	24.80	11.80	3.62	0.68
200	—	—	—	30.70	14.50	4.48	0.84
225	—	—	—	38.80	18.40	5.60	1.06
250	—	—	—	47.90	22.70	7.00	1.31
275	—	—	—	—	27.40	7.70	1.59
300	—	—	—	—	32.60	10.10	1.88

For SI: 1 inch = 25.4 mm, 1 pound per square inch = 6.895 kpa, 1 gallon per minute = 3.785 L/m.

APPENDIX E

TABLE E103.3(5)
ALLOWANCE IN EQUIVALENT LENGTHS OF PIPE FOR FRICTION LOSS IN VALVES AND THREADED FITTINGS (feet)

FITTING OR VALVE	PIPE SIZE (inches)							
	1/2	3/4	1	1 1/4	1 1/2	2	2 1/2	3
45-degree elbow	1.2	1.5	1.8	2.4	3.0	4.0	5.0	6.0
90-degree elbow	2.0	2.5	3.0	4.0	5.0	7.0	8.0	10.0
Tee, run	0.6	0.8	0.9	1.2	1.5	2.0	2.5	3.0
Tee, branch	3.0	4.0	5.0	6.0	7.0	10.0	12.0	15.0
Gate valve	0.4	0.5	0.6	0.8	1.0	1.3	1.6	2.0
Balancing valve	0.8	1.1	1.5	1.9	2.2	3.0	3.7	4.5
Plug-type cock	0.8	1.1	1.5	1.9	2.2	3.0	3.7	4.5
Check valve, swing	5.6	8.4	11.2	14.0	16.8	22.4	28.0	33.6
Globe valve	15.0	20.0	25.0	35.0	45.0	55.0	65.0	80.0
Angle valve	8.0	12.0	15.0	18.0	22.0	28.0	34.0	40.0

For SI: 1 inch = 25.4 mm, 1 foot = 304.8 mm, 1 degree = 0.0175 rad.

TABLE E103.3(6)
PRESSURE LOSS IN FITTINGS AND VALVES EXPRESSED AS EQUIVALENT LENGTH OF TUBE[a] (feet)

NOMINAL OR STANDARD SIZE (inches)	FITTINGS				Coupling	VALVES			
	Standard Ell		90-Degree Tee			Ball	Gate	Butterfly	Check
	90 Degree	45 Degree	Side Branch	Straight Run					
3/8	0.5	—	1.5	—	—	—	—	—	1.5
1/2	1	0.5	2	—	—	—	—	—	2
5/8	1.5	0.5	2	—	—	—	—	—	2.5
3/4	2	0.5	3	—	—	—	—	—	3
1	2.5	1	4.5	—	—	0.5	—	—	4.5
1 1/4	3	1	5.5	0.5	0.5	0.5	—	—	5.5
1 1/2	4	1.5	7	0.5	0.5	0.5	—	—	6.5
2	5.5	2	9	0.5	0.5	0.5	0.5	7.5	9
2 1/2	7	2.5	12	0.5	0.5	—	1	10	11.5
3	9	3.5	15	1	1	—	1.5	15.5	14.5
3 1/2	9	3.5	14	1	1	—	2	—	12.5
4	12.5	5	21	1	1	—	2	16	18.5
5	16	6	27	1.5	1.5	—	3	11.5	23.5
6	19	7	34	2	2	—	3.5	13.5	26.5
8	29	11	50	3	3	—	5	12.5	39

For SI: 1 inch = 25.4 mm, 1 foot = 304.8 mm, 1 degree = 0.01745 rad.

a. Allowances are for streamlined soldered fittings and recessed threaded fittings. For threaded fittings, double the allowances shown in the table. The equivalent lengths presented above are based on a C factor of 150 in the Hazen-Williams friction loss formula. The lengths shown are rounded to the nearest half-foot.

APPENDIX E

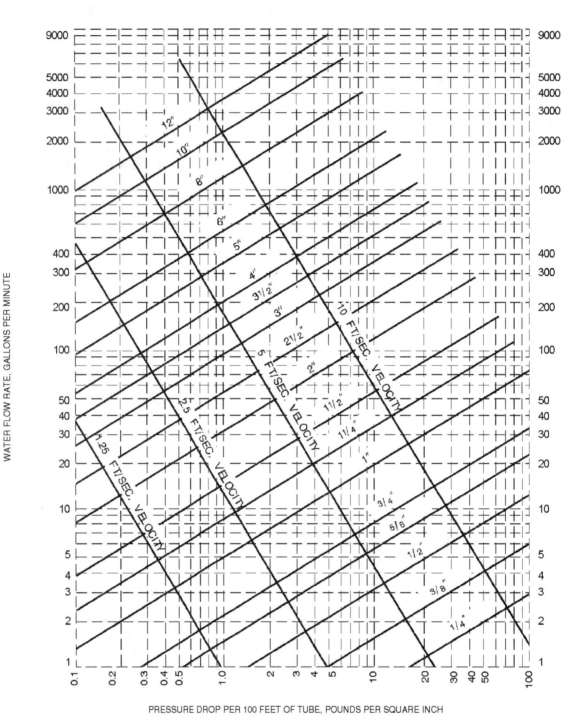

PRESSURE DROP PER 100 FEET OF TUBE, POUNDS PER SQUARE INCH

Note: Fluid velocities in excess of 5 to 8 feet/second are not usually recommended.

For SI: 1 inch = 25.4 mm, 1 foot = 304.8 mm, 1 gpm = 3.785 L/m, 1 psi = 6.895 kPa, 1 foot per second = 0.305 m/s.

a. This chart applies to smooth new copper tubing with recessed (streamline) soldered joints and to the actual sizes of types indicated on the diagram.

FIGURE E103.3(2)
FRICTION LOSS IN SMOOTH PIPE[a] (TYPE K, ASTM B88 COPPER TUBING)

APPENDIX E

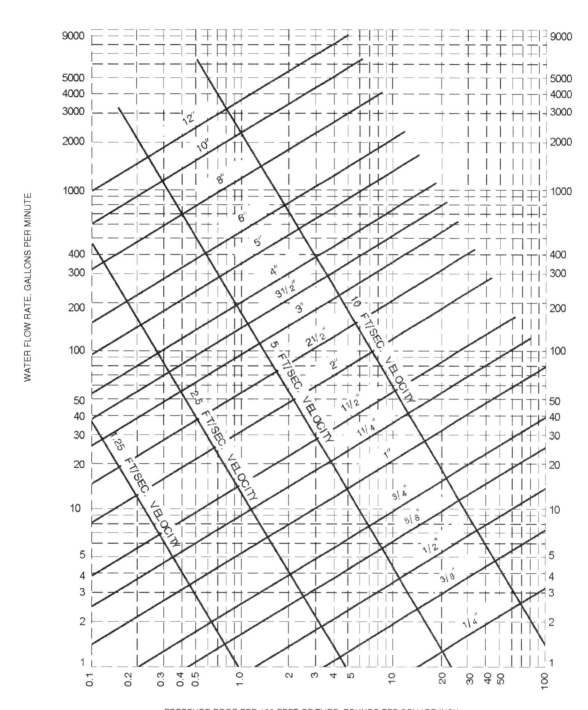

PRESSURE DROP PER 100 FEET OF TUBE, POUNDS PER SQUARE INCH

Note: Fluid velocities in excess of 5 to 8 feet/second are not usually recommended.

For SI: 1 inch = 25.4 mm, 1 foot = 304.8 mm, 1 gpm = 3.785 L/m, 1 psi = 6.895 kPa, 1 foot per second = 0.305 m/s.

a. This chart applies to smooth new copper tubing with recessed (streamline) soldered joints and to the actual sizes of types indicated on the diagram.

FIGURE E103.3(3)
FRICTION LOSS IN SMOOTH PIPE[a] (TYPE L, ASTM B88 COPPER TUBING)

APPENDIX E

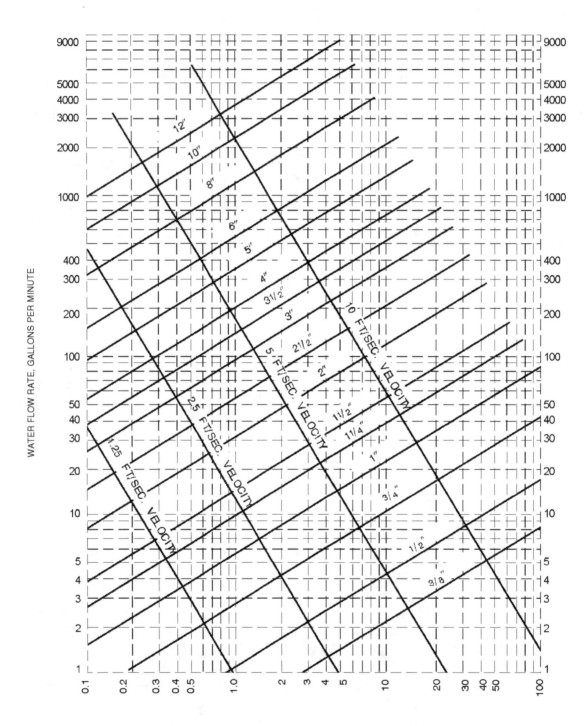

PRESSURE DROP PER 100 FEET OF TUBE, POUNDS PER SQUARE INCH

Note: Fluid velocities in excess of 5 to 8 feet/second are not usually recommended.

For SI: 1 inch = 25.4 mm, 1 foot = 304.8 mm, 1 gpm = 3.785 L/m, 1 psi = 6.895 kPa, 1 foot per second = 0.305 m/s.
a. This chart applies to smooth new copper tubing with recessed (streamline) soldered joints and to the actual sizes of types indicated on the diagram.

FIGURE E103.3(4)
FRICTION LOSS IN SMOOTH PIPE[a] (TYPE M, ASTM B88 COPPER TUBING)

APPENDIX E

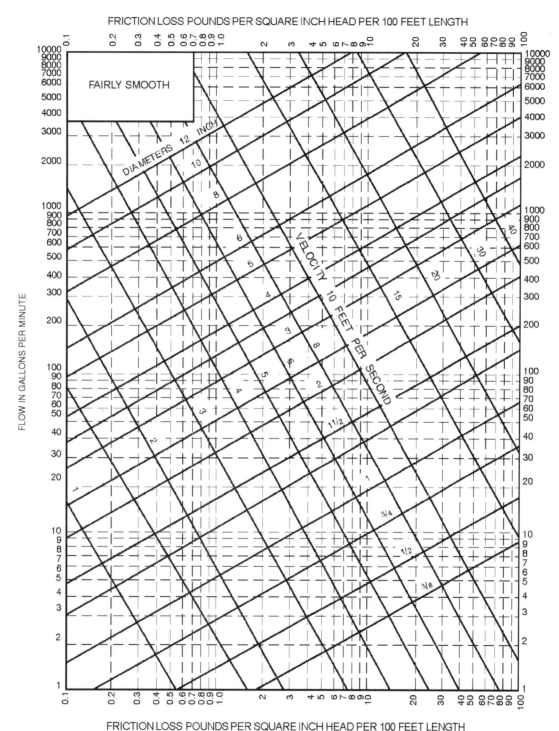

For SI: 1 inch = 25.4 mm, 1 foot = 304.8 mm, 1 gpm = 3.785 L/m, 1 psi = 6.895 kPa, 1 foot per second = 0.305 m/s.
a. This chart applies to smooth new steel (fairly smooth) pipe and to actual diameters of standard-weight pipe.

**FIGURE E103.3(5)
FRICTION LOSS IN FAIRLY SMOOTH PIPE**[a]

APPENDIX E

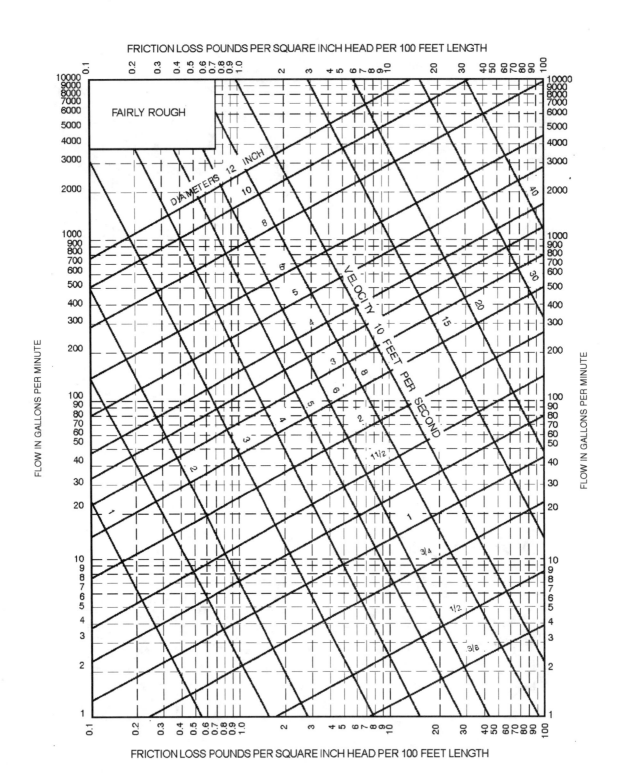

For SI:1 inch = 25.4 mm, 1 foot = 304.8 mm, 1 gpm = 3.785 L/m, 1 psi = 6.895 kPa, 1 foot per second = 0.305 m/s.
a. This chart applies to fairly rough pipe and to actual diameters, which in general will be less than the actual diameters of the new pipe of the same kind.

**FIGURE E103.3(6)
FRICTION LOSS IN FAIRLY ROUGH PIPE[a]**

APPENDIX E

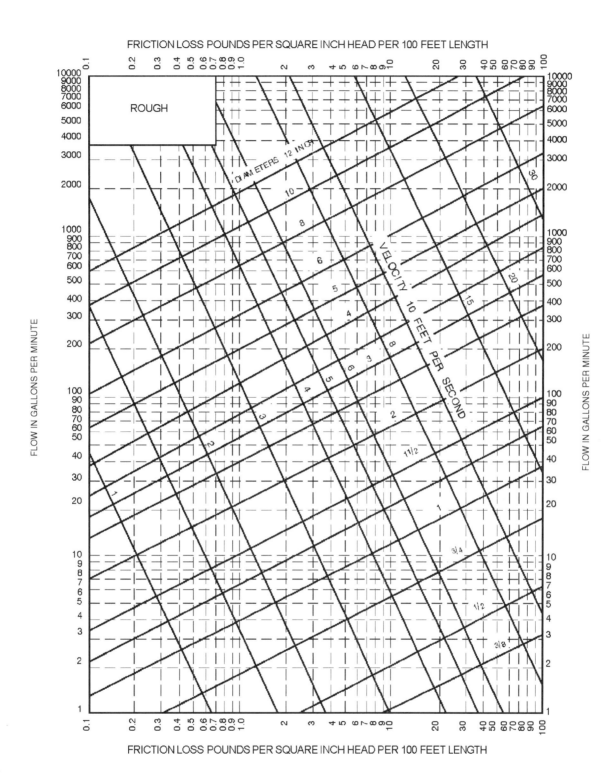

For SI: 1 inch = 25.4 mm, 1 foot = 304.8 mm, 1 gpm = 3.785 L/m, 1 psi = 6.895 kPa, 1 foot per second = 0.305 m/s.
a. This chart applies to very rough pipe and existing pipe and to their actual diameters.

**FIGURE E103.3(7)
FRICTION LOSS IN ROUGH PIPE[a]**

APPENDIX E

SECTION E201
SELECTION OF PIPE SIZE

E201.1 Size of water service mains, branch mains and risers. The minimum size water service pipe shall be $^3/_4$ inch (19.1 mm). The size of water service mains, *branch* mains and risers shall be determined according to water supply demand [gpm (L/m)], available water pressure [psi (kPa)] and friction loss due to the water meter and *developed length* of pipe [feet (m)], including equivalent length of fittings. The size of each water distribution system shall be determined according to the procedure outlined in this section or by other design methods conforming to acceptable engineering practice and *approved* by the code official:

1. Supply load in the building water distribution system shall be determined by total load on the pipe being sized, in terms of water-supply fixture units (w.s.f.u.), as shown in Table E103.3(2). For fixtures not listed, choose a w.s.f.u. value of a fixture with similar flow characteristics.

2. Obtain the minimum daily static service pressure [psi (kPa)] available (as determined by the local water authority) at the water meter or other source of supply at the installation location. Adjust this minimum daily static pressure [psi (kPa)] for the following conditions:

    2.1. Determine the difference in elevation between the source of supply and the highest water supply outlet. Where the highest water supply outlet is located above the source of supply, deduct 0.5 psi (3.4 kPa) for each foot (0.3 m) of difference in elevation. Where the highest water supply outlet is located below the source of supply, add 0.5 psi (3.4 kPa) for each foot (0.3 m) of difference in elevation.

    2.2. Where a water pressure-reducing valve is installed in the water distribution system, the minimum daily static water pressure available is 80 percent of the minimum daily static water pressure at the source of supply or the set pressure downstream of the pressure-reducing valve, whichever is smaller.

    2.3. Deduct all pressure losses due to special equipment such as a backflow preventer, water filter and water softener. Pressure loss data for each piece of equipment shall be obtained through the manufacturer of such devices.

    2.4. Deduct the pressure in excess of 8 psi (55 kPa) due to installation of the special plumbing fixture, such as temperature controlled shower and flushometer tank water closet. Using the resulting minimum available pressure, find the corresponding pressure range in Table E201.1.

3. The maximum *developed length* for water piping is the actual length of pipe between the source of supply and the most remote fixture, including either hot (through the water heater) or cold water branches multiplied by a factor of 1.2 to compensate for pressure loss through fittings. Select the appropriate column in Table E201.1 equal to or greater than the calculated maximum *developed length*.

4. To determine the size of water service pipe, meter and main distribution pipe to the building using the appropriate table, follow down the selected "maximum *developed length*" column to a fixture unit equal to, or greater than the total installation demand calculated by using the "combined" water supply fixture unit column of Table E103.3(2). Read the water service pipe and meter sizes in the first left-hand column and the main distribution pipe to the building in the second left-hand column on the same row.

5. To determine the size of each water distribution pipe, start at the most remote outlet on each *branch* (either hot or cold *branch*) and, working back toward the main distribution pipe to the building, add up the water supply fixture unit demand passing through each segment of the distribution system using the related hot or cold column of Table E103.3(2). Knowing demand, the size of each segment shall be read from the second left-hand column of the same table and maximum *developed length* column selected in Steps 1 and 2, under the same or next smaller size meter row. The size of any *branch* or main does not need to be larger than the size of the main distribution pipe to the building established in Step 4.

SECTION E202
DETERMINATION OF PIPE VOLUMES

E202.1 Determining volume of piping systems. Where required for engineering design purposes, Table E202.1 shall be used to determine the approximate internal volume of water distribution piping.

APPENDIX E

TABLE E201.1
MINIMUM SIZE OF WATER METERS, MAINS AND DISTRIBUTION PIPING
BASED ON WATER SUPPLY FIXTURE UNIT VALUES (w.s.f.u.)

METER AND SERVICE PIPE (inches)	DISTRIBUTION PIPE (inches)	MAXIMUM DEVELOPMENT LENGTH (feet)									
Pressure Range 30 to 39 psi		40	60	80	100	150	200	250	300	400	500
3/4	1/2 a	2.5	2	1.5	1.5	1	1	0.5	0.5	0	0
3/4	3/4	9.5	7.5	6	5.5	4	3.5	3	2.5	2	1.5
3/4	1	32	25	20	16.5	11	9	7.8	6.5	5.5	4.5
1	1	32	32	27	21	13.5	10	8	7	5.5	5
3/4	1 1/4	32	32	32	32	30	24	20	17	13	10.5
1	1 1/4	80	80	70	61	45	34	27	22	16	12
1 1/2	1 1/4	80	80	80	75	54	40	31	25	17.5	13
1	1 1/2	87	87	87	87	84	73	64	56	45	36
1 1/2	1 1/2	151	151	151	151	117	92	79	69	54	43
2	1 1/2	151	151	151	151	128	99	83	72	56	45
1	2	87	87	87	87	87	87	87	87	87	86
1 1/2	2	275	275	275	275	258	223	196	174	144	122
2	2	365	365	365	365	318	266	229	201	160	134
2	2 1/2	533	533	533	533	533	495	448	409	353	311

METER AND SERVICE PIPE (inches)	DISTRIBUTION PIPE (inches)	MAXIMUM DEVELOPMENT LENGTH (feet)									
Pressure Range 40 to 49 psi		40	60	80	100	150	200	250	300	400	500
3/4	1/2 a	3	2.5	2	1.5	1.5	1	1	0.5	0.5	0.5
3/4	3/4	9.5	9.5	8.5	7	5.5	4.5	3.5	3	2.5	2
3/4	1	32	32	32	26	18	13.5	10.5	9	7.5	6
1	1	32	32	32	32	21	15	11.5	9.5	7.5	6.5
3/4	1 1/4	32	32	32	32	32	32	32	27	21	16.5
1	1 1/4	80	80	80	80	65	52	42	35	26	20
1 1/2	1 1/4	80	80	80	80	75	59	48	39	28	21
1	1 1/2	87	87	87	87	87	87	87	78	65	55
1 1/2	1 1/2	151	151	151	151	151	130	109	93	75	63
2	1 1/2	151	151	151	151	151	139	115	98	77	64
1	2	87	87	87	87	87	87	87	87	87	87
1 1/2	2	275	275	275	275	275	275	264	238	198	169
2	2	365	365	365	365	365	349	304	270	220	185
2	2 1/2	533	533	533	533	533	533	533	528	456	403

(continued)

TABLE E201.1—continued
MINIMUM SIZE OF WATER METERS, MAINS AND DISTRIBUTION PIPING BASED ON WATER SUPPLY FIXTURE UNIT VALUES (w.s.f.u.)

METER AND SERVICE PIPE (inches)	DISTRIBUTION PIPE (inches)	MAXIMUM DEVELOPMENT LENGTH (feet)									
Pressure Range 50 to 60 psi		40	60	80	100	150	200	250	300	400	500
3/4	1/2 [a]	3	3	2.5	2	1.5	1	1	1	0.5	0.5
3/4	3/4	9.5	9.5	9.5	8.5	6.5	5	4.5	4	3	2.5
3/4	1	32	32	32	32	25	18.5	14.5	12	9.5	8
1	1	32	32	32	32	30	22	16.5	13	10	8
3/4	1 1/4	32	32	32	32	32	32	32	32	29	24
1	1 1/4	80	80	80	80	80	68	57	48	35	28
1 1/2	1 1/4	80	80	80	80	80	75	63	53	39	29
1	1 1/2	87	87	87	87	87	87	87	87	82	70
1 1/2	1 1/2	151	151	151	151	151	151	139	120	94	79
2	1 1/2	151	151	151	151	151	151	146	126	97	81
1	2	87	87	87	87	87	87	87	87	87	87
1 1/2	2	275	275	275	275	275	275	275	275	247	213
2	2	365	365	365	365	365	365	365	329	272	232
2	2 1/2	533	533	533	533	533	533	533	533	533	486

METER AND SERVICE PIPE (inches)	DISTRIBUTION PIPE (inches)	MAXIMUM DEVELOPMENT LENGTH (feet)									
Pressure Range Over 60		40	60	80	100	150	200	250	300	400	500
3/4	1/2 [a]	3	3	3	2.5	2	1.5	1.5	1	1	0.5
3/4	3/4	9.5	9.5	9.5	9.5	7.5	6	5	4.5	3.5	3
3/4	1	32	32	32	32	32	24	19.5	15.5	11.5	9.5
1	1	32	32	32	32	32	28	28	17	12	9.5
3/4	1 1/4	32	32	32	32	32	32	32	32	32	30
1	1 1/4	80	80	80	80	80	80	69	60	46	36
1 1/2	1 1/4	80	80	80	80	80	80	76	65	50	38
1	1 1/2	87	87	87	87	87	87	87	87	87	84
1 1/2	1 1/2	151	151	151	151	151	151	151	144	114	94
2	1 1/2	151	151	151	151	151	151	151	151	118	97
1	2	87	87	87	87	87	87	87	87	87	87
1 1/2	2	275	275	275	275	275	275	275	275	275	252
2	2	365	368	368	368	368	368	368	368	318	273
2	2 1/2	533	533	533	533	533	533	533	533	533	533

For SI: 1 inch = 25.4, 1 foot = 304.8 mm.

a. Minimum size for building supply is 3/4-inch pipe.

APPENDIX E

TABLE E202.1
INTERNAL VOLUME OF VARIOUS WATER DISTRIBUTION TUBING

	OUNCES OF WATER PER FOOT OF TUBE								
Size Nominal, Inch	Copper Type M	Copper Type L	Copper Type K	CPVC CTS SDR 11	CPVC SCH 40	CPVC SCH 80	PE-RT SDR 9	Composite ASTM F 1281	PEX CTS SDR 9
3/8	1.06	0.97	0.84	N/A	1.17	—	0.64	0.63	0.64
1/2	1.69	1.55	1.45	1.25	1.89	1.46	1.18	1.31	1.18
3/4	3.43	3.22	2.90	2.67	3.38	2.74	2.35	3.39	2.35
1	5.81	5.49	5.17	4.43	5.53	4.57	3.91	5.56	3.91
1 1/4	8.70	8.36	8.09	6.61	9.66	8.24	5.81	8.49	5.81
1 1/2	12.18	11.83	11.45	9.22	13.20	11.38	8.09	13.88	8.09
2	21.08	20.58	20.04	15.79	21.88	19.11	13.86	21.48	13.86

For SI: 1 ounce = 0.030 liter.

INDEX

A

ABS PIPE
   Approved standards . . . . Table 605.3,
          Table 605.5, Table 702.1, Table 702.2,
                  Table 702.3, Table 702.4,
              Table 1102.4, Table 1102.7
   Support. . . . . .Table 308.5

ACCESS TO
   Air admittance valve .918.5
   Appliances 502.5
   Backflow preventers 608.14
   Backwater valve 714.3
   Cleanouts 708.1.10, 1103.4
   Filters 1302.5, 1302.9, 1303.8, 1303.12
   First-flush diverter. .1303.4
   Fixtures. 405.2
   Flexible water connectors. 605.6
   Floor drain 413.2
   Flush tank components 415.3.4
   Grease removal devices. 1003.3.6
   Interceptors and separators 1003.5, 1003.10
   Manifolds 604.10.3
   Pumps 602.3.5, 1302.9
   Reservoir or interior 1301.9.6
   Shower compartment 421.4.2
   Slip joints 405.9
   Standpipes 802.4.3
   Sump 712.3.2
   Vacuum breakers 415.2
   Vacuum station. 713.4
   Valves. 503.1, 604.11, 606.3, 712.2,
        715.5, 918.5, 1302.8.1, 1302.9, 1303.12
   Waste receptors 802.4
   Water heaters 502.3, 502.5
   Whirlpool pump. 426.5

ACCESSIBLE 202

ACCESSIBLE PLUMBING FACILITIES
   Clearances 404.2, 405.3.1
   Protection required 404.3
   Route 403.5
   Signs. 403.4
   Where required. 404.1

ADMINISTRATION
   Applicability. 102
   Approval 105, 106.5.7, 107.5
   Inspections 107

   Intent. 101.3, 109.1
   Maintenance 102.3
   Means of appeal 109
   Scope 101.2
   Validity 106.5.2
   Violations 108

AIR ADMITTANCE VALVE
   Chemical waste. 901.3
   Definition. 202
   Where permitted 905.1, 917.6, 918

AIR BREAK
   Definition. 202
   Method of providing 802.1.5, 802.1.6,
                                        802.1.7
   Where required 406.2, 802.1.2, 802.1.5, 802.3.2

AIR GAP
   Annual inspection of 312.10.1
   Application of . . . . Table 608.1
   Bidet 408.2
   Clothes washer. 406.1
   Definition. 202
   Dishwasher. 409.2
   Method of providing 608.16.1, 802.3.1
   Required 416.4, 417.1, 504.6, 606.5.6, 608.3.1,
       608.4, 608.10, 608.14.1, 608.14.2, 608.14.2.1,
       608.14.3, 608.14.8, 608.15.2.1, 608.17.1.1,
          608.17.1.2, 608.17.2, 608.17.3,
       608.17.4.1, 608.17.10, 609.7, 611.2, 802.1.1,
          802.1.2, 802.1.3, 802.1.4, 802.1.5,
           802.1.6, 802.1.7, 802.3, 802.3.1

AIR TEST. 312.3, 312.5, 312.8

ALTERATIONS AND REPAIRS 102.2.1,
                              102.4, 307.4, 612.1

ALTERNATE ON-SITE NONPOTABLE WATER
   Definition. 202
   Systems 1302

ALTERNATIVE ENGINEERED DESIGN
   Definition. 202
   Requirements for 316
   Special inspections of. 107.3

ALTERNATIVE MATERIALS AND
EQUIPMENT. 105.2, 605.1

APPROVED
   Definition. 202

ATMOSPHERIC VACUUM BREAKER 412.8,
                Table 608.1, 608.3.1, 608.14.4,
                608.14.6, 608.16.4, 608.16.4.1,
              608.16.4.2, 608.17.4.1, 608.17.5

INDEX

AREA DRAIN (See STORM DRAIN)
AUTOMATIC CLOTHES WASHER 301.3,
Table 403.1, 406, 413.4,
608.16.4.2, Table 709.1,
802.4.3.1, 1003.6, 1302.2, 1402

B

BACKFILLING 107.2, 306, 1403.1.5
BACKFLOW PREVENTER
  Access to 608.15
  Boiler 608.17.2
  Carbonated beverage dispenser 608.17.1.1
  Coffee and noncarbonated
    beverage dispenser 608.17.1.2
  Definition 202
  Effect of installation 607.3
  Fire sprinkler system 608.17.4
  Humidifier 608.17.10
  Lawn irrigation 608.17.5, 1301.11,
    1302.11, 1303.14. 1304.3.1
  Location of 608.15
  Relief port piping from 608.15.2.1
  Required 406.1, 408.2, 409.2, 416.4, 417.1, 608.1
  Standards . . . . Table 608.1, 608.14.2,
    608.14.3. 608.14.9
  Testing . . . 312.10.2, 1302.12.4, 1303.15.6. 1304.4.2
BACKFLOW PROTECTION 608.1
BACKWATER VALVE
  Where required 714, 802.1.2, 1101.8,
    1101.9, 1111.1, 1302.8.1,
    1302.8.2, 1303.11.1
BAROMETRIC LOOP . . . . Table 608.1, 608.14.4
BATHING ROOM 403.1.2
BATHTUB
  Discharge from 301.3, 1302.2
  Doors in 407.4
  Drainage fixture unit 709.1
  Faucet for 412
  Foot (pedicure bathtub) 423.3
  Glazing nearby 407.3
  Outlet and overflow for 407.2
  Recessed area for trap below slab 1002.8
  Required . . . . Table 403.1
  Standards 407.1
  Walls around built-in type 421.4.1
  Water temperature supplied to 412.5, 412.7
  Whirlpool 202, 426
BEVERAGE DISPENSER 608.17.1
BIDET
  Backflow protection 408.2
  Standards 408.1
  Water temperature supplied to 408.3
BOOSTER SYSTEM 604.7, 606.5,
606.5.1. 606.5.5
BOTTLING ESTABLISHMENT 1003.7
BRACING, SWAY 308.6
BRASS PIPE
  (See COPPER OR COPPER ALLOY PIPE)
BUILDING DRAIN
  Branches of . . . . Table 710.1(2)
  Cleanouts for 708.1.1, 708.1.3, 708.1.4
  Connection to building sewer 705.16.4
  Definition 202
  Existing 703.4
  Material, above-ground . . . . Table 702.1
  Material, below ground . . . . Table 702.2
  Sizing . . . . Table 710.1(1)
BUILDING DRAINAGE SYSTEM
  Below sewer level 712.1
  Definition (See DRAINAGE SYSTEM) 202
  Excluding detrimental materials from 302.1
  Oil Separators for 1003.4
BUILDING SEWER
  Cleanouts 708.1.2, 708.1.4
  Definition 202
  In same trench as nonpotable
    water piping 1301.11
  In same trench as storm sewer 703.3
  Connection to building drain 708.1.3
  Material . . . . Table 702.3
  Reuse of existing 703.4
  Separation from water service 603.2
  Sizing . . . . Table 710.1(1)
  Testing 312
BUILDING TRAP
  Definition 202
  Prohibition of 1002.6
BUNDLED WATER PIPING 606.7

C

CAST-IRON PIPE
  Approved standards . . . . . . Table 702.1, Table 702.2,
    Table 702.3, Table 702.4,
    Table 1102.4, Table 1102.5,
    Table 1102.7
  Joints 705.3. 705.16.1. 705.16.3
  Support of . . . . Table 308.5
CAULKING FERRULES 705.18
CHANGE IN DIRECTION OF DRAINAGE
  Anchorage for 308.7
  Cleanouts for 708.1.4
  Support against sway 308.6

CHEMICAL WASTE
Exclusion from the sewer 803.1
Neutralizing. 803.1
Requirements 702.6
System venting 901.3
CIRCUIT VENT
Connection of 905.1
Definition. 202
Requirements for system 914
Sizing 906.2
CIRCULATING HOT WATER SYSTEM
Definition. 202
Required. 607.2
CLAY PIPE
Approved standards . . . . Table 702.3, Table 1102.4, Table 1102.5
Joints 705.11
CLEANOUT
Building drain 708.1.1
Change of direction 708.1.4
Clearances 708.1.9
Condensate drain 314.2.5
Definition. 202
Direction of flow 708.1.8
Floor accessed 708.1.10.2
Horizontal drains. 708.1.1
Location of 708.1.10
Manholes 708.1.1, 708.1.2, 708.1.7
On stacks 708.1.5
Plugs, materials for. 708.1.6
Prohibited use of. 708.1.11
Sewers 708.1.2
Size. 708.1.5
Trim covers for plugs 708.1.10.1
Where required. 708.1, 708.2, 708.3, 708.4
CLEARANCES
. . . . .404.2, 405.3.1, 502.5, 918.8
CLEAR WATER WASTE
. . . 802.1.5, 803.1.3, 803.1.4
CLOTHES WASHER
(See AUTOMATIC CLOTHES WASHER)
CODE OFFICIAL
Application for permit .106.3
Appointment 103.2
Approval of modifications 105
Definition. 202
Department records 104.7
Duties and powers 104
General. 104.1
Identification 104.5
Inspections 104.3, 107
COFFEE MACHINE
. . . . .608.16.1.2

COLLECTION PIPE
Definition. 202
On-site nonpotable water reuse system 1302.4
Testing of 1302.12.1, 1303.15.3
COMBINATION WASTE AND VENT SYSTEM
Definition. 202
Requirements 915
COMBINED BUILDING DRAIN
. . . . 202
COMMON VENT
. . . . 202, 911
COMPARTMENT, WATER CLOSET
. . . . 405.3.1, 405.3.4
CONCRETE PIPE
Approved standards . . . . . Table 702.3, Table 1102.4
CONDENSATE DRAIN
Configured for clearing blockages 314.2.5
Piping material 314.2.2
Mini-split systems 314.2.4.1
Traps. 314.2.4
CONDUCTOR
Definition. 202
CONFLICTS
. . . . 301.7
CONNECTION TO PUBLIC SEWER
. . . . 701.2
CONNECTIONS TO SUMPS AND EJECTORS
Below sewer level 712.1
CONSERVATION, WATER AND ENERGY
Fixtures 604.4
Flow rates . . . . Table 604.3
Insulation. 505, 607.2.1, 607.5
CONSTRUCTION DOCUMENTS
. . . . 106.3.1
CONTROLLED FLOW STORM DRAINAGE
. . . . 1110
COPPER OR COPPER-ALLOY PIPE OR TUBING
Approved standards . . . . . Table 605.3, Table 605.4, Table 702.1, Table 702.2, Table 702.3
COPPER SHEET (See SHEET COPPER)
CORROSION PROTECTION OF PIPING
. . . . 305.1, 605.1
CPVC PIPE OR TUBING
Approved standards . . . . . . Table 605.3, Table 605.4
CROSS CONNECTIONS
. . . . 608
CUTTING OR NOTCHING, STRUCTURAL MEMBERS
. . . . 307.2, 307.4, Appendix C

D

DEMAND RECIRCULATION WATER SYSTEM
Controls. 607.2.1.2
Definition. 202
DESIGN PROFESSIONAL
. . . . 107.3.2, 109.2.1, 307.4, 316.1

INDEX

DETRIMENTAL WASTES 302, 1003.1
DISHWASHING MACHINE 409, 802.1.6
DISINFECTION OF POTABLE
   WATER SYSTEM 610
DISTANCE OF TRAP FROM VENT . . . . . . Table 909.1
DISTRIBUTION SYSTEM (See WATER)
DRAIN
   Roof 1105
   Storm 1106
DRAINAGE FIXTURE UNITS
   Definition 202
   Values for continuous flow. 709.3
   Values for fixtures . . . . Table 709.1
DRAINAGE SYSTEM
   Connection to sewer or private
      disposal system 701.2
   Determining size 710.1
   Fixture units . . . . Table 709.1, 709.3
   Indirect waste. 802
   Joints 705
   Material detrimental to 302
   Materials 702
   Obstructions. 706.2
   Offset sizing. 711
   Provisions for future fixtures 710.2
   Sizing. . . . . Table 710.1(1), Table 710.1(2)
   Slope of piping. . . . . Table 704.1
   Sumps and ejectors 712
   Testing. 312
DRINKING FOUNTAIN
   Accessibility requirements 410.3
   Approvals. 410.1
   Definition 202
   Location 403.5
   Number required . . . . Table 403.1, 410.1
   Prohibited location 410.5
   Small occupancies. 410.2
   Substitution for. 410.4
DRINKING WATER TREATMENT UNIT 611

E

EJECTOR, SEWAGE 712
EJECTOR CONNECTION 712.3.5
ELEVATOR SHAFT 301.6
EMERGENCY SHOWER 411
EMPLOYEE TOILET FACILITIES 403.3
ENGINEERED DESIGN, ALTERNATIVE 316
EXCAVATION. 306
EXISTING BUILDING 102.2.1
EXISTING BUILDING
   PLUMBING SYSTEMS 102.2, 102.2.1
EYEWASH STATION 411

F

FACILITIES, TOILET 403
FAMILY AND ASSISTED USE 403.1.2, 403.2.1
FAUCETS 412
FEES. 106.6
FERRULES (See CAULKING FERRULES)
FILL VALVE 415.3.1
FITTINGS
   Approved standards. . . . . . . Table 605.5, Table 702.4
   Drainage system 706
FIXTURE CALCULATIONS 403.1.1
FIXTURE FITTINGS 412
FIXTURE LOCATION 405.3.1
FIXTURE TRAPS 1002
   Acid-resisting 1002.9
   Building 1002.6
   Design of 1002.2
   For each fixture 1002.1
   Prohibited. 1002.3
   Seals 1002.4
   Seal protection 1002.4.1
   Setting and protection 1002.7
   Size 1002.5
FIXTURE UNITS, DRAINAGE
   Definition 202
   Values for continuous flow 709.3
   Values for fixtures . . . . 709.1, Table 709.1
FIXTURES (See PLUMBING FIXTURES)
FLOOD LEVEL RIM 202
FLOOD-RESISTANT CONSTRUCTION
   Design flood elevation 202
   Flood hazard area 202
   Flood hazard resistance. 309
FLOOR DRAINS . . . . 412, Table 709.1
FLOOR DRAINS, EMERGENCY . . . . 202, Table 709.1
FLOOR FLANGES 405.4.1, 405.4.2
FLOW RATES . . . . 604.3, Table 604.3,
                                Table 604.4, 604.10
FLUSHING DEVICES
   Dual 425.1
   Flush tanks. . . . . 415.2, Table 604.3, Table 604.5
   Flushometer tanks 415.2, 604.3, 604.5
   Flushometer valves . . . . 415.2, Table 604.3,
                                       Table 604.5
   Required. 415
FOOD HANDLING DRAINAGE 802.1.1, 802.1.7
FOOD WASTE DISPOSER 416
FOOTBATHS (PEDICURE BATHS) 423.3
FOOTINGS, PROTECTION OF 307.5
FREEZING, PROTECTION OF
   BACKFLOW DEVICES 608.15.2

INDEX

FREEZING, PROTECTION OF PIPES 305.4
FROST CLOSURE 903.2
FULL OPEN VALVE
   Definition................................202
   Required................................605.3
FUTURE FIXTURES 704.4

G

GALVANIZED STEEL PIPE .... Table 605.3, Table 605.4, 605.17
GARBAGE CAN WASHERS 417
GARBAGE DISPOSALS
  (See FOOD WASTE DISPOSER)
GENERAL REGULATIONS 301
   Conflicts 301.7
   Connection to plumbing system 301.3
   Connection to public water and sewer .. 301.3, 301.4
   Elevator machinery rooms 301.6
   Health and safety ...108.7, 108.7.1, 108.7.2, 108.7.3
   Materials detrimental 302, 302.1, 302.2
   Piping sizes indicated.....................301.5
   Protection of pipes 305
   Rodentproofing..... 304, 304.1, 304.2, 304.3, 304.4
   Sleeves................................305.3
   Strains and stresses in pipe 305.2
   Toilet facilities for workers 311
   Trenching, excavation and backfill 306
   Washroom requirements 310
GRAYWATER (See also ALTERNATE
  ON-SITE NONPOTABLE WATER)
   Color of distribution piping 608.9.2.1
   Definition................................202
   Flushing 1302.6.1
   Subsurface irrigation....................1402.1
   Trap priming 1002.2.4.1
GREASE INTERCEPTORS................. 202, 1003
   Additives to........................... 1003.3.3
   Approved standards 1003.3.5, 1003.3.7
   Capacity .... Table 1003.3.5.1, 1003.4.2.1, 1003.4.2.2
   Discharge 1003.3.8
   Food waste disposers to................ 1003.3.2
   Gravity-type 1003.3.7
   Hydromechanical-type 1003.3.5
   Not required 1003.3.4
   Piping leading to.........................704.1
   Required 1003.3.1
GUTTERS .... 1106.6, Table 1106.6

H

HANGERS AND SUPPORTS
   Attachment to buildings................... 308.4
   Definitions................................202
   Material............................... 308.3
   Seismic............................... 308.2
   Spacing......................... Table 308.5
HEALTH AND SAFETY................... 101.3
HEALTH CARE PLUMBING 609
HEAT EXCHANGERS...................608.17.3
HORIZONTAL PIPE
   Definition................................202
   Slope for drainage........................ 704.1
HOT WATER
   Circulating system.................. 202, 607.2.1
   Definition................................202
   Demand recirculation water system 607.2.1.2
   Flow of hot water from fixtures 607.4
   Heaters and tanks.................... Chapter 5
   Pipe insulation......................... 607.5
   Pumps for water heaters..................607.2.1
   Recirculating systems with
     thermostatic mixing valves 607.2.2
   Supply system........................... 607
   Temperature maintenance (Heat trace) 607.2.1
HOUSE TRAP (See BUILDING TRAP)
HUMIDIFIER 608.17.10

I

IDENTIFICATION OF
  BUNDLED WATER PIPING................ 606.7
INDIRECT WASTE 802
   Air gap or break 802.3.1, 802.3.2
   Food handling establishment 802.1.1
   Receptors 802.4
   Special wastes 803
   Wastewater temperature.................. 702.5
   Where required......................... 802.1
INDIVIDUAL VENT 910
INSPECTION............................ 107.1
   Final 107.2
   Reinspection...........................107.4.3
   Required 107.2
   Rough-in 107.2
   Scheduling of 107.2
   Testing 107.4
INSPECTOR (See CODE OFFICIAL)
INSULATION 505, 607.5

INTERCEPTORS AND SEPARATORS 1003
   Access to 1003.10
   Additives to 1003.3.3
   Approved standards. 1003.3.5, 1003.3.7
   Bottling establishments 1003.7
   Capacity of grease interceptors . . . . Table 1003.3.4.1
   Clothes washers 1003.6
   Connection of discharge 1003.3.8
   Definitions 202
   Fats, oils and greases systems 1003.3.7
   Not required 1003.3.4
   Oil and flammable liquids separators 1003.4
   Rate of flow control for
      grease interceptors. 1003.3.5.2
   Required 1003.1
   Slaughterhouses 1003.8
   Venting. 905.4, 1003.9
IRRIGATION, LAWN 608.17.5
ISLAND FIXTURE VENT 916

J

JOINTS AND CONNECTIONS 605, 705
   ABS plastic pipe 605.10, 705.2
   Between different materials 605.23, 705.16
   Brazed joints 605.12.1, 605.13.1,
                                705.5.1, 705.6.1
   Building drain to building sewer
      (ABS/PVC) 705.16.4
   Cast-iron pipe 705.3
   Caulked joint 705.3.1, 705.7.1
   Concrete pipe 705.4
   Copper or copper-alloy pipe 605.12, 705.5
   Copper or copper-alloy tubing 605.13, 705.6
   CPVC plastic pipe 605.14, 605.15
   Expansion joints 308.8
   Flared 605.13.2, 605.16.1, 605.18.1
   Galvanized steel pipe 605.17
   Grooved and shouldered 605.13.3, 605.17.3,
                              605.21.2, 605.22.3
   Heat fusion 605.18.2, 605.19.1,
                         705.12.1, 705.13.1, 705.14.1
   Mechanical joints 605, 705
   Polyethylene of
      raised temperature (PE-RT) 605.24
   Polyethylene plastic pipe or tubing (PE) 605.18
   Polypropylene plastic pipe or tubing (PP) 605.19
   Press-connect 202, 605.13.5
   Prohibited 605.9, 707
   Push-fit 605.13.7, 605.14.4, 605.16.3
   PVC plastic pipe 605.21, 705.10
   Screwed together 605.23.1, 705.16.2
   Slip joints 405.9, 1002.2
   Soldered joints 605.12.3, 605.13.6,
                                 705.5.3, 705.6.3
   Solvent cemented 605.10.2, 605.14.2,
                              605.15.2, 605.21.2,
                      705.2.2, 705.10.2, 705.16.4
   Stainless steel 605.22
   Threaded 605.10.3, 605.12.4,
                     605.14.3, 605.17.1, 605.21.4,
                       705.2.3, 705.5.4, 705.8.1,
                    705.10.3, 705.16.2, 705.16.3
   Vitrified clay pipe 705.11
   Welded 605.12.5, 605.22.2, 705.5.5

K

KITCHEN
   Grease interceptor 1003.3.1
   Hot water for 609.3
   Island sink 916.1
   Passage through 403.3.1
   Sink drainage fixture unit . . . . Table 709.1
   Sink requirement . . . . Table 403.1
   Sink standards 422

L

LABELS FOR BUNDLED WATER PIPING 606.7
LAUNDRIES (See CLOTHES WASHERS)
LAUNDRY TRAY 418
LAUNDRY TUB (See LAUNDRY TRAY)
LAVATORY 403.1.3, 419
LEADER
   Connecting to combined sewer 1103
   Definition 202
   For rainwater collection systems 1303.3
   Sizing . . . . 1106.1, 1106.3, Table 1106.3
   Slope 1303.5.1
LIGHT AND VENTILATION REQUIREMENT . . . 310.1
LOADING, SANITARY
   DRAINAGE SYSTEM 709
LOCATION
   Anchorage for drain piping 308.7.1
   Drinking fountain 403.5, 410.5
   Fixture 405.3
   Fixtures, obstruction 405.3.3
   Potable water supply tank 606.5.8
   Prohibited, drinking fountain 410.5
   Secondary roof drain discharge 1108.2
   Storage tank 1302.7.1,
                      1303.10.1, Table 1303.10.1
   Subsoil irrigation site 1402.3
   Toilet facilities, malls 403.3.4

Toilet facilities, other than in malls 403.3.3
Toilet room and kitchen 403.3.2
Trench 307.5
Valve. 606.1, 606.2
Vent terminal 903.5
Water heater. 501.4
Wells. . . . . 608.18.8, Table 1302.7.1

M

MANHOLE. 708.1.1, 708.1.2, 708.1.7,
714.1, 1301.9.6
MANIFOLD . . . . Table 604.5, 604.10
MATERIAL
  Above-ground drainage and
    vent pipe . . . . .Table 702.1
  Alternative. 105.2
  Approved 105.4, 107.2.3, 308.3, 316,
402.1, 504.7, 608.18.5,
702.6. 708.1.6, 712.3.2,
1002.2, 1002.3, 1113.1.2, 1303.2
  Building sewer pipe 702.3
  Building storm sewer pipe 1102.4
  Chemical waste system 702.6
  Fittings . . . . .Table 605.5, Table 702.4,
Table 1102.7
  Identification 303.1
  Joint 605, 705
  Roof drain 1102.6
  Sewer pipe 702.3
  Standards for . . . . Chapter 15
  Storm sewer pipe . . . . .Table 1102.4
  Subsoil drain pipe. 1102.5
  Underground building
    drainage and vent pipe 702.2
  Vent pipe 702.1, 702.2
  Water distribution pipe . . . . .Table 605.4
  Water service pipe . . . . .Table 605.3
MATERIAL, FIXTURE
  Quality 402.1
  Special use. 402.2
MATERIAL, SPECIAL
  Caulking ferrules. 705.18
  Cleanout plugs 708.1.6
  Sheet copper 402.3
  Sheet lead 402.4
  Soldering bushings. 705.19
MECHANICAL JOINTS. 605, 705
MEDICAL GAS, NONFLAMMABLE 1202.1

N

**NONPOTABLE WATER
REUSE SYSTEMS** 1302
NONPOTABLE WATER SYSTEMS
  Identification 608.9
  Disinfection 1301.2.1
  Distribution piping color. 608.9.2.1
  Filtration 1301.2.2, 1302.5, 1303.8
  Protection of potable water from 608.1, 1301.5
  Requirements for . . . . Chapter 13
  Signage. 1301.3
  Tanks 1301.9

O

OFFSET
  Closet flange. 704.2
  Definition, piping 202
  Drainage 704.3, 710.1.1, 710.1.2, 711
  Venting 907, 917.7
OPENINGS
  Through walls or roofs 304.4, 305.5, 315
OVERFLOW 405.8, 407.2, 415.3.1,
415.3.2, 606.5.4, 1101.7, 1108.1
OXYGEN SYSTEMS 1203.1

P

PARKING GARAGE 305.7, 403.3,
1002.1, 1003.4.2.2
PAN
  Auxiliary (condensate) 314.2.3
  Water heater. 504.7
PARTITION
  Fixture. 405.3.1, 405.3.4, 405.3.5
  Trap. 1002.2, 1002.3
PEDICURE BATHS 423.3
PERMIT
  Application for. 106.3
  Conditions of. 106.5
  Fees 106.6
  Posting of 106.5.8
  Suspension of 106.5.5
PENETRATIONS. 307.3, 315
PIPE BURSTING 716
PIPING
  Construction documents 104.2, 106.3.1
  Drainage, horizontal slope 704.1
  Drainage piping installation 704
  Drainage piping offset, size. 710.1.1,
710.1.2, 711.2

Joints 605, 705
   Rehabilitation of the inside of. 601.5
PIPING PROTECTION
   Backfilling. 306.3
   Bedding 306.1, 306.2
   Corrosion. 305.1
   Expansion and contraction 305.2
   Exposed locations 305.6
   Foundation wall 305.3
   Freezing. 305.4
   Impact by vehicles 305.7
   Installation 305.2
   Penetration by fasteners 305.6
   Slabs 305.1
   Structural settlement 305.2
   Tunneling. 306.4
PLUMBING FACILITIES 403
PLUMBING FIXTURES
   Accessible 404
   Automatic clothes washers 406
   Bidet. 408
   Clearances. 404.2, 405.3.1, 405.3.5
   Definition 202
   Dishwashing machine 409
   Drainage fixture unit values . . . . Table 709.1
   Drinking fountain 410
   Emergency showers 411
   Eyewash stations. 411
   Floor and trench drains 413
   Floor sinks 414
   Food waste disposer 416
   Future fixtures 704.4
   Garbage can washer 414.7
   Installation of 405
   Joints at wall or floor 405.6
   Kitchen sink 422
   Laundry tray. 418
   Laundry tub (See Laundry tray)
   Lavatories 419
   Materials 402.1
   Minimum number of 403.1
   Ornamental pools 423.1
   Quality of 402.1
   Separate facilities for 403.2
   Setting 405.3
   Showers. 421
   Sinks 422
   Urinals 424
   Water closets 425
   Water coolers 202, 410
   Water supply protection 608.2

PLUMBING INSPECTOR (See CODE OFFICIAL)
PNEUMATIC (SEWAGE) EJECTORS 712
POLYETHYLENE PIPE OR TUBING
   PE approved standards . . . . Table 702.2,
                               Table 1102.4, Table 1102.5
   PE-AL-PE approved standards . . . . Table 605.3,
                                         Table 605.4
   PEX approved standards . . . . Table 605.3,
                                         Table 605.4
   PEX-AL-PEX approved standards . . . . . Table 605.3,
                                         Table 605.4
   (PE-RT) approved standards . . . . Table 605.3,
                                         Table 605.4
POLYPROPYLENE (PP) PIPE OR TUBING
   Approved standards . . . . . . Table 605.3, Table 605.4,
                                         Table 702.3
POTABLE WATER, PROTECTION OF 608
POTABLE WATER
   HANDLING EQUIPMENT 608.4
PRESSURE OF WATER
   DISTRIBUTION SYSTEM 604.6, 604.7, 604.8
PROHIBITED
   Air admittance valve use 918.3.2, 918.8
   Bends in tubing 605.18.4
   Building traps 1002.6
   Cleanout opening use 708.1.11
   Conductor use 1104
   Cross connections 608.7
   Discharge to water reuse system. 1302.2.1
   Drinking fountain location. 410.5
   Fixture design. 401.2
   Joints and connections. 605.9, 707
   Offset in waste stack 913.2
   Piping reutilization 608.10
   Plumbing systems location. 301.6
   Stack connections 917.8
   Storm water into sanitary sewer. 1101.3
   Tank location 606.5.8
   Toilet facility location 403.3.2
   Trap design 1002.3
   Valve location. 608.8
   Vent terminal use 903.4
   Water cooler location 410.5
   Water dispenser locations 410.5
   Well pits 602.3.5.1
   Yard hydrant. 608.8
PROTECTION OF POTABLE WATER 603.2, 608
PUBLIC SWIMMIMG POOL . . . . 202, Table 403.1
PUMPED WASTE FIXTURE 405.5
PUMPING EQUIPMENT 608.17.9, 712,
                                         1113, 1303.12

PUSH-FIT JOINT (See CONNECTIONS AND JOINTS)
PVC PIPE . . . . .314.2.2, Table 605.3, Table 605.4,
Table 702.1, Table 702.2, Table 702.3,
703.2, 712.3.3.1, Table 1102.4,
Table 1102.5, Table 1102.7,Table 1403.2

Q

QUALITY OF WATER
  General 602.3.3, 608.1, 1002.4.1.2,
1301.2, 1301.9.2, 1302.6, 1302.12.6,
1303.5, 1303.9, 1303.15.8, 1303.15.9
QUICK CLOSING VALVE
  Causing water hammer .604.9
  Definition 202

R

RAINWATER COLLECTION AND
  DISTRIBUTION SYSTEMS 1303
RAINWATER
  (STORM WATER) DRAINAGE . . . . Chapter 11
RAINWATER QUAILITY 1303.15.9
RECLAIMED WATER SYSTEMS 1304
REDUCED PRESSURE PRINCIPLE
BACKFLOW PREVENTER
  Boiler potable water supply 608.17.2
  Definition 202
  Freeze protection of 608.15.2
  Installation of 608.14.2
  Outdoor enclosures for 606.15.1
  Periodic inspections 312.10
  Relief port piping from 608.15.2.1
  Standards . . . . .Table 608.1
  Where required 608.17.2, 608.17.4,
608.17.4.1, 608.17.5, 608.17.6

REGISTERED DESIGN PROFESSIONAL
  (See DESIGN PROFESSIONAL)
REFERENCED STANDARDS . . . . Chapter 15
RELIEF VALVE, PRESSURE 504.4, 504.5
  Definition 202
RELIEF VALVE DISCHARGE PIPING 504.6
RELIEF VALVE, VACUUM (See VACUUM)
RELIEF VENT 905.5, 906.2, 906.3, 908, 914.4,
914.4.2, 918.3.1
  Definition 202
ROOF DRAIN
  Definition 202
  Flow rating for water height above . . . 1101.7, 1105.2

  Installation of 1105.1
  Standards for 1106.2
ROUGH-IN INSPECTION 107.2

S

SANITARY SEWER (See BUILDING SEWER)
SANITARY TEE . . . . 706.3, Table 706.3, 917.3
SCREWED JOINTS
  (See JOINTS AND CONNECTIONS)
SCUPPERS 1106.5, 1108.1
SEPARATE (TOILET) FACILITIES 403.2
SERVICE SINKS . . . . Table 403.1, 422,
604.4, Table 709.1
SEWER (See BUILDING SEWER)
SHEET COPPER 305.5, 402.3, 415.3,
421.5.2.4, 902.2
SHEET LEAD 402.4, 421.5.2.3, 902.3
SHAMPOO SINKS 412.10
SHOWER LINER
  Material 421.5.2
  Testing 312.9
SHOWERS
  Approvals for prefabricated types 421.1
  Compartment size 421.4
  Emergency 411
  Required number 403.1
  Wall protection for 421.4.1
  Water supply riser for 421.2
SILL COCK . . . . 412.6, Table 604.3, 604.8,
606.2, 608.16.4.2, 1301.12
SINGLE-STACK VENT SYSTEM 917
SINKS . . . . Table 403.1, 414, 422
SIPHONIC ROOF DRAINAGE SYSTEMS 1107
SIZING
  Drainage system 710
  Fixture drain 709
  Fixture water supply 604.5
  Vent system 906
  Water distribution system 604
  Water service 603.1
SLAUGHTERHOUSES 1003.8
SLEEVES IN FOUNDATION WALLS 305.3
SLIP JOINTS (See JOINTS AND CONNECTIONS)
SLOPE, PIPING . . . . Table 709.1
SOLAR HEATING SYSTEM 502.1
SOLDERING BUSHINGS 705.19

INDEX

SOLVENT CEMENTING
 (See JOINTS AND CONNECTIONS)
SPECIALTY PLUMBING FIXTURES........... 423
SPECIAL WASTES....................... 803
STAINLESS STEEL DRAINAGE SYSTEMS
 Approved standards for
  drainage systems................. Table 702.1,
   Table 702.2, Table 702.3,
   Table 702.4, Table 1102.4,
   Table 1102.5, Table 1102.7
 Joints for drainage systems........ 705.16.7, 705.20
STAINLESS STEEL PIPE
 Approved standards for
  water systems........ Table 605.3, Table 605.4
 Joints for..................... 605.22, 605.23
STANDARDS (See REFERENCED STANDARDS)
STANDPIPE (FIXTURE).................... 802.4
STEEL PIPE (See also MATERIAL)
 Galvanized, approved standards....... Table 605.4
 Stainless, approved standards....... Table 605.3,
   Table 605.4
STORM DRAIN
 Area drain definition....................... 202
 Building size....................... 1106.1
 Building subdrains..................... 1112.1
 Building subsoil drains................... 1111.1
 Conductors and connections............. 1104
 Definition............................... 202
 General............................. 1101
 Inspection of......................... 107.2
 Prohibited drainage.................... 1101.3
 Roof drains................... 1102.6, 1105
 Secondary roof drains................... 1108
 Sizing of conductors, leaders
  and storm drains................... 1106
 Sizing of roof gutters................. 1106.6
 Sizing of vertical conductors and leaders.... 1106.2
 Traps............................ 1103
 Where required................... 1101.2
STRAPS (See HANGERS AND SUPPORTS)
STRUCTURAL SAFETY......... 307, Appendix C
SUBDRAIN BUILDING................... 1112.1
SUBSOIL DRAIN PIPE................... 1111.1
SUBSOIL LANDSCAPE
 IRRIGATION SYSTEMS.............Chapter 14
SUMP VENT........................ 906.5
SUMP PUMP DISCHARGE PIPE...... 301.6, 712.3.3
SUMPS........................... 1113
SUPPORTS (See also PIPING PROTECTION)... 308
SWIMMING POOL
 Definition............................... 202
 Public................... 202, Table 403.1

 Solar heating of...................... 612.1
 Waste connections................. 801
 Water connections to................ 423.1
SWIMMING POOL DRAINAGE............ 802.1.4

T

TEMPERATURE-ACTUATED
 FLOW REDUCTION VALVE............. 412.7
TEMPERATURE AND
 PRESSURE RELIEF VALVE............. 504
TEMPERATURE CONTROL
 Mixing valves.......... 411.3, 412.3, 412.4, 412.5,
  Table 604.3, 607.2.2, 607.4, 613.1
TEMPERATURE OF WASTEWATER........ 702.5
TEMPERED WATER.... 202, 419.5, 607.1.1, 607.1.2
TEST
 Drainage and vent air test................. 312.3
 Drainage and vent final test............. 312.4
 Drainage and vent water test............ 312.2
 Forced sewer test.................. 312.7
 Gravity sewer test................... 312.6
 Percolation test.................. 1402.2
 Required tests.................... 312.1
 Shower liner..................... 312.9
 Test gauges..................... 312.1.1
 Test of backflow prevention devices....... 312.10
 Test of conductors.................. 312.8
THERMAL EXPANSION CONTROL........ 607.3
THERMAL EXPANSION TANK, SUPPORT... 308.10
THIRD-PARTY CERTIFICATION........ 202, 303.3,
  303.4, 303.5, 501.5,
  605.3, 605.15.2, 705.10.2
THREADED JOINTS
 (See JOINTS AND CONNECTIONS)
TOILET FACILITIES
 Definition............................... 202
 Required........................... 403
 Signs.............................. 403.4
 Single-user........................ 403.1.2
 Travel distance............... 403.3.3, 403.3.4
 Workers............................. 311
TOILET ROOM DOOR LOCKING......... 403.3.6
TOILETS (See WATER CLOSETS)
TRAP SEAL PROTECTION............. 1002.4.1
TRAPS
 Acid-resisting..................... 1002.9
 Building......................... 1002.6
 Design........................... 1002.2
 Prohibited types................... 1002.3
 Seal............................ 1002.4
 Separate for each fixture............ 1002.1
 Size................... Table 709.1, Table 709.2

INDEX

TRENCH DRAINS 413
TRENCHING, EXCAVATION AND BACKFILL . . . 306
TUNNELING 306.4

U

**UNDERGROUND DRAINAGE
AND VENT PIPE** 702.2
URINAL PARTITIONS 405.3.5
URINALS . . . . Table 403.1, 424

V

VACUUM BREAKER 202, 312.10.2, 412.8,
415.2, Table 608.1, 608.3.1,
608.14.5, 608.14.6, 608.16.4,
608.16.4.1, 608.16.4.2, 609.4
VACUUM RELIEF VALVE 504.1, 504.2
VACUUM DRAINAGE SYSTEMS 715
VACUUM SYSTEM, MEDICAL 1202.1
VALVE
  Air admittance 202, 918
  Approved standards . . . . Table 605.7, 714.2
  Backwater 202, 714
  Full-open 202, 605.3
  Pressure reducing 604.8
  Relief 202, 504.4
  Stop and waste 608.8
  Temperature-actuated flow
    reduction 412.7
  Transfer valves 412.8
VENT
  Individual 910
  Sizing 906
VENT, RELIEF 908
VENT STACK
  Definition 202
  Required 904.2
VENTS AND VENTING
  Branch vents 906.4
  Circuit 914
  Combination waste 915
  Common 911
  Definition 202
  Distance from trap . . . . Table 909.1, 909.2
  Engineered systems 919
  Island fixture 916
  Required vent 904.1
  Single stack 917
  Vent stack 904.2
  Vent terminal 903
  Waste stack 913
  Wet vent 912

W

WALL-HUNG WATER CLOSET 425
**WASHING MACHINE
  (See AUTOMATIC CLOTHES WASHER
  and DISHWASHER)**
WASTE
  Indirect 802
  Special 803
WASTE RECEPTORS
  Definition 202
WASTE STACK 913
WASTEWATER TEMPERATURE 702.5
WATER
  Cross connection 608.7
  Distribution system design 604
  Distribution piping material . . . . Table 605.4
  Excessive pressure 604.8
  Hammer 604.9
  Pressure booster 606.5
  Rain 202
  Reclaimed 202
  Service 603
  Service piping material . . . . Table 605.3
  Storm 202
WATER CLOSETS 425
WATER CLOSET COMPARTMENTS 405.3.4
**WATER CLOSET PERSONAL
  HYGIENE DEVICE** 412.9
WATER CONSERVATION
  Fixtures 604.4
  Flow rates . . . . Table 604.4
WATER COOLER
  Approvals 410.1
  Definition 202
  Prohibited location 410.5
WATER DISPENSER
  Definition 202
  Prohibited location 410.5
  Substitution for drinking fountain 410.4
WATER HAMMER 604.9
WATER HEATER
  Definition 202
  Requirements for . . . . Chapter 5
WATER HEATER PAN 504.7
WATER TREATMENT
  Drinking water systems 611
  Softeners 608.4
WELL WATER 602.3.1
WET VENT 912
WIPED JOINTS 705.9.2
WORKERS' TOILET FACILITIES 311

EDITORIAL CHANGES – THIRD PRINTING

Page 30, **Section 408.2:** lines 3 and 4 now read . . . venter in accordance with Section 608.14.1, 608.14.2, 608.14.3, 608.14.5 or 608.14.6.

Page 56, **Section 608.17.7:** line 5 now reads . . . 608.14.6 or 608.14.8.

Page 71, **Section 802.1:** line 5 now reads . . . fied in Sections 802.1.1 through 802.1.7. Fixtures not

Page 80, **Section 919.1:** line 3 now reads . . . and testing requirements of Section 316.

Page 87, **Section 1101.9:** line 3 now reads . . . drainage systems in accordance with Section 714.

Page 111, new Reference Standard has been added to **ANSI**. It reads . . .

A118.10—99: Specifications for Load Bearing, Bonded, Waterproof Membranes for Thin Set Ceramic Tile and Dimension Stone Installation
                421.5.2.5, 421.5.2.6

Page 126, Reference Standard **TCNA** has been deleted.

For the complete errata history of this code, please visit: https://www.iccsafe.org/errata-central/

INTERNATIONAL CODE COUNCIL®

People Helping People Build a Safer World

BENEFITS THAT WORK FOR YOU

No matter where you are in your building career, put the benefits of ICC Membership to work for you!

Membership in ICC connects you to exclusive I-Codes resources, continuing education opportunities and *Members-Only* benefits that include:

- Free code opinions from I-Codes experts
- Free I-Code book(s) or download to new Members*
- Discounts on I-Code resources, training and certification renewal
- Posting resumes and job search openings through the ICC Career Center
- Mentoring programs and valuable networking opportunities at ICC events
- Free benefits — Corporate and Governmental Members: Your staff can receive free ICC benefits too*
- *Savings of up to 25% on code books and training materials and more*

Put the benefits of ICC Membership to work for you and your career. **Visit www.iccsafe.org/mem3 to join now or to renew your Membership.** Or call 1-888-ICC-SAFE (422-7233), ext. 33804 to learn more today!

* Some restrictions apply. Speak with an ICC Member Services Representative for details.

ICC EVALUATION SERVICE

In Cooperation with **Innovation RESEARCH LABS**

Look no further than ICC-ES PMG® The Leading provider of product evaluations in plumbing, mechanical and fuel gas, with excellent customer service and the highest acceptability by code officials at the price you're looking for.

Benefits of having an ICC-ES PMG Listing

- ICC-ES PMG offers a lower cost for certification than competitors
- Expedited certification for all client listings
- ICC-ES PMG does not conduct warehouse inspections
- ICC-ES PMG does not charge for additional company listings
- ICC-ES will accept test reports from other entities
- No fee for EPA WaterSense listings and lead law listings
- No separate file for NSF 61 listings
- ANSI and SCC Accredited

www.icc-es-pmg.org
800-423-6587 x7643

People Helping People Build a Safer World

Looking for the missing piece?
Solve the puzzle and advance your career with ICC University

ICC University has been built from the ground up with you in mind.

Take advantage of tools to help you better track and manage your career growth and professional development, including automatic CEU tracking and simplified search options to find code training. Plan for your future and manage your professional development from one, easy-to-use location.

ICC University provides you with:

- Simplified access to over 300 training options
- Automatic CEU tracking to keep you on track for recertification
- Robust curriculums that identify supporting courses, publications and exam study materials to assist you in preparing for certification exams and achieving your next professional milestone
- The ability to purchase all courses, related publications and exam preparation materials – as well as register for certification exams – from a single screen
- And more!

www.iccsafe.org/ExploreICCU

17-14099

 **ICC INTERNATIONAL CODE COUNCIL**

People Helping People Build a Safer World®

Valuable Guides to Changes in the 2018 I-Codes®

NEW!

FULL COLOR! HUNDREDS OF PHOTOS AND ILLUSTRATIONS!

SIGNIFICANT CHANGES TO THE 2018 INTERNATIONAL CODES®

Practical resources that offer a comprehensive analysis of the critical changes made between the 2015 and 2018 editions of the codes. Authored by ICC code experts, these useful tools are "must-have" guides to the many important changes in the 2018 International Codes.

Key changes are identified then followed by in-depth, expert discussion of how the change affects real world application. A full-color photo, table or illustration is included for each change to further clarify application.

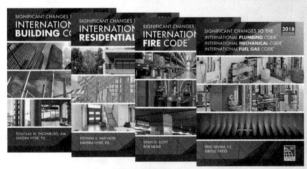

SIGNIFICANT CHANGES TO THE IBC, 2018 EDITION
#7024S18

SIGNIFICANT CHANGES TO THE IRC, 2018 EDITION
#7101S18

SIGNIFICANT CHANGES TO THE IFC, 2018 EDITION
#7404S18

SIGNIFICANT CHANGES TO THE IPC/IMC/IFGC, 2018 EDITION
#7202S18

ORDER YOUR HELPFUL GUIDES TODAY!
1-800-786-4452 | www.iccsafe.org/books

HIRE ICC TO TEACH
Want your group to learn the Significant Changes to the I-Codes from an ICC expert instructor? Schedule a seminar today!

email: **ICCTraining@iccsafe.org** | phone: **1-888-422-7233 ext. 33818**